Thank you, Madagascar

About the author

ALISON JOLLY (May 9, 1937–February 6, 2014) was a primatologist known for her studies of lemur biology. She wrote for both popular and scientific audiences and conducted extensive fieldwork on lemurs in Madagascar, primarily at the Berenty Reserve, a small private reserve of gallery forest set in the semi-arid spiny desert area in the far south of Madagascar.

Thank you, Madagascar

The conservation diaries of Alison Jolly

ALISON JOLLY

Zed Books
LONDON

To all the people of Madagascar
who work for the good of their country.

And especially to my friend and colleague
Dr Hantanirina Rasamimanana

Thank you, Madagascar: The Conservation Diaries of Alison Jolly was first published in 2015.

This edition was published in 2016.

Published by Zed Books Ltd,
The Foundry, 17 Oval Way, London SE11 5RR, UK.

www.zedbooks.co.uk

Designed and typeset by illuminati, Grosmont

Index by John Barker

A catalogue record for this book is available from the British Library.

ISBN 978-1-78360-318-3 hb
ISBN 978-1-78360-317-6 pb
ISBN 978-1-78360-319-0 pdf
ISBN 978-1-78360-320-6 epub
ISBN 978-1-78360-321-3 mobi

Printed and bound by TJ International Ltd, Padstow, Cornwall

Contents

Acknowledgements

For their very helpful comments on the manuscript, I thank David Attenborough, Keith Bezanson, Melanie Dammhahn, Ann Downer-Hazell, Lee Durrell, Nick Fairclough, François Falloux, David Foster, Jörg Ganzhorn, Lisa Gaylord, Jacques Gérin, Jane Goodall, Frans Lanting, Mia-Lana Lührs, Bernhard Meier, Russ Mittermeier, Emma Napper, Dai Peters, Joe Peters, Peter Porter Lowry III, Johny Rabenantoandro, Léon Rajaobelina, Ny Fanja Rakotomalala, Joelisoa Ratsirarson, Alison Richard, Takayo Soma, Eleanor Sterling, Manon Vincelette, Pat Wright. Also to Hantanirina Rasamimanana and the support she has been given by the École Normale Supérieure and by her precious family, her husband Niry Ratovonirina and sons Andou, Zo and Tsilavina. And all the photographers who kindly sent photos, particularly Cyril Ruoso for the magical cover image. I also thank Andrea Cornwall for her help in suggesting the inspirational Zed Books, and Lucy at illuminati.

And then I thank my wonderfully supportive family: my husband of fifty years Richard Jolly, and my four children, Margaretta, Susan, Arthur and Richard Brabazon Jolly.

Foreword

Alison Jolly was known in the academic world for her ground-breaking work as a primatologist, but somehow that title feels wrong. It conjures up the image of a soulless scientist, whilst Alison was one of the warmest, funniest and most passionate women that you could meet. More than any other person she was instrumental in initiating me, so by extension the thousands who read my guide to Madagascar, into a 38–year love affair with the island in all its diversity and complexity.

I first came to Madagascar as a tourist in 1976, with zero knowledge of the wildlife. When I saw my first lemurs in Nosy Komba I thought the sexually dimorphic black lemurs (only the males are black) were two different species. By that time Alison had been studying ring-tailed lemurs in Berenty for fourteen years and was *the* expert on the subject. So when I was asked to lead a pioneering tour of Madagascar in 1982 I bought her book *A World Like Our Own: Man and Nature in Madagascar.* It changed my life. Here was a description of all aspects of this lovely, but challenging country, with intimate portraits of lemurs but also of the people and the dilemmas of promoting conservation in an island where poverty is rife. In that book I also 'met' Richard Jolly in the dedication. 'Tell the whole story,' he said; 'ecology with people, not just your animals.' And that's what she did, with Richard's continuing encouragement, for the rest of her life. Lemurs were only part of the picture, not the obsessive whole, because she knew and understood the people—from dignitaries to peasants—as well as she knew the lemurs. *A World Like Our Own* showcased Alison the writer. Her talent for narrative and description is the equal of the very best of our travel writers, and brought the island of Madagascar to the notice of the general public for the first time. For a while we had a lively correspondence about the possibility of my reissuing the book as a paperback, which sadly never happened, but it gave me an excuse to get to know her and she was a generous contributor to many editions of my guidebooks.

Alison was surely the funniest primatologist ever; not, I think, through any conscious effort, but because that's how she was. Her humour was infectious. You might start a serious discussion about lemur behavior but end up hooting with laughter over the lighter side of Madagascar. Also Alison was as anthropomorphic about lemurs as the rest of us. When the albino lemur Sapphire (subject of a TV programme) died, she told me 'The death of Little Nell was nothing compared with our reactions to the demise of this little lemur.' Where she was absolutely serious, however, was when discussing conservation issues where her views were her own and based on her intimate knowledge of the country, rather than popular but less informed, opinion. Thus she came down firmly on the side of the controversial Rio Tinto titanium mine—and I can't imagine that anyone listening to her arguments could have disagreed with her. As she said: 'If you think that people and forest will somehow muddle through before the hills are scraped as bare as Haiti, then there is no reason to think that money and organization will improve life. If you look at the statistics of forest loss, you opt for the mine.'

Perhaps her most accessible book of all was *Lords and Lemurs: Mad Scientists, Kings with Spears, and the Survival of Diversity in Madagascar.* It is Alison at her best: funny, fascinating and illuminating. Anyone who has been to, or is thinking of going to, Berenty can enjoy it.

So we come to her last book, dictated during the final months of her life, but based on the diaries she wrote during all those changing decades in Madagascar. And as you read you understand that she wrote because she had to. Her work in Madagascar threw up so many triumphs, frustrations, joys and disappointments that a natural writer like Alison *needed* to record them. And how lucky we are that she could do it so well, and that Zed Books is bringing these unique accounts to all those who love Madagascar, whether tourist, nature lover, dedicated conservationist or professional primatologist. What a legacy!

Hilary Bradt

INTRODUCTION

My adventurous and astonishing mother

Alison Jolly overturned established thinking after becoming the first scientist to do an in-depth field study of the behaviour of the ring-tailed lemur, *L. catta*, beginning work in Madagascar in 1962 as a young graduate from Yale. She discovered that this species—and as it turned out almost all other lemurs—have female dominance over males, breaking the then orthodoxy that primates were male-dominant. As she later joked, the 'king' of the DreamWorks animation *Madagascar* ought to have been a queen.

My mother saw the ring-tails as pugnacious, swaggering, but also formal to the point of ritual and ever so maternally doting.[1] She herself was not pugnacious at all, though she did indeed dote upon me and her three other children. Nor did she swagger, though as a nearly six foot American in her characteristically loud shirts, beads and sneakers, she cut a very visible figure in Lewes, the small English town in which she made her home. She was a gentle woman, with a lyrical, sing-song way of speaking, who on the rare occasions when hurt or cross would retreat into a detective novel with a glass of sherry (or whisky if very out of sorts), and a bowl of salted cashews. She saw the world as a place where humans

1. A. Jolly, *Lords and Lemurs: Mad Scientists, Kings with Spears, and the Survival of Diversity in Madagascar* (2004), pp. 30–31.

could be cruel but wonderfully amusing, and where animals, plants, trees and seas, would always be more magical than any human invention.

Such a gentle sense of wonder is one of the reasons this book exists. Based on diaries written over the decades of her visits to Madagascar, it shows her hope that even such a complex and impoverished country could achieve a balanced ecology, although, like evolution itself, it might be slow and painful. This belief, and her love of her fragrant research site at Berenty, kept her writing. What follows is a unique insider's account of a major conservation effort in one of the world's most iconic biodiversity hotspots. But these diaries are also powered by her pleasure in storytelling, where accounts of policy-making and science conferencing, community discussion and conservation camping, were also farces, romances and natural comedy, even when they expressed political tragedy.

Her sometimes mischievously literary eye came from her upbringing. Growing up in the hilly university town of Ithaca, New York, she was the only child of the artist Alison Mason Kingsbury and the humorist and Cornell scholar Morris Bishop. Neither of her parents was remotely left-wing, but they were sophisticates with wonderful taste and humanity. East Coast Europhile Americans, their favourite writers included Dante and Proust, and they always dined with side plates and silver. Her father, whose light verse was regularly published in the *New Yorker*, taught her to observe 'truth with laughter, not with tears,'[2] her mother to see beauty and order in the landscape. Mason Kingsbury, a sometime muralist for New York's Radio City Music Hall who was also supported by the Federal Art Project in the 1930s, proved that a woman could be as professional as a man.[3]

2. M. Bishop, *A Bowl of Bishop: Museum Thoughts, and Other Verses* (1954), Preface.
3. Both Alison's parents' papers are at the Cornell University Library as the Morris

Her mother's model proved to be the one that my mother adopted, combining mothering and wifehood with a largely self-directed career, in part made possible by an inheritance from her grandfather Albert Kingsbury, who had established a lucrative engineering business in 1912. But in many other ways, Alison broke with her parents' genteel Anglo-Europeanism. Choosing science, choosing Madagascar, she threw off the corset and girdle, though not all of the side plates and silver, to join a set of brilliant women who came of age in the 1960s as the pioneers in the new field of primatology. Mum's work can be set alongside that of Jane Goodall,[4] Dian Fossey and Biruté Galdikas, who have made the lives of chimpanzees, gorillas and orang-utans so vivid to us. Although she was not sponsored by Louis Leakey, she did publish with *National Geographic*, and as this book shows, she worked with many legendary scientists and activists, including her supervisor G. Evelyn Hutchinson, Gerald and Lee Durrell, David Pilbeam, Jerome Bruner, Jean-Jacques Petter, Ian Tattersall, Russell Mittermeier, Tom Lovejoy, Sarah Hrdy, Patricia Wright, Hirai Hirohisa, Naoki Koyama, her one-time student Alison Richard—as well as outstanding Malagasy scientist pioneers, including Berthe Rakotosamimanana, Joseph Andriamampianina, Joel Ratsirason, Barthélémy Vaohita, Gilbert Ravelojaona, Guy Ramanantsoa, Philibert Tsimamandro and her closest colleague Hanta Rasamimanana. She could also schmooze with the best when it came to the World Wildlife Fund or the World Bank, not to mention various ministers and presidents of Madagascar. But she did not take a full-time academic post, and she chose to study prosimians, rather than great apes, a more distant set of relatives to humankind.

Bishop papers and the Alison Mason Kingsbury papers. Her mother's artistic career is analyzed in J. Piccirilli, *The Art and Life of Alison Mason Kingsbury* (2010).

4. Mum heard Jane give her first scientific paper at a special symposium on primates of the London Zoological Society in 1962 and was deeply impressed.

Less compelling to most of us perhaps, for Mum, lemurs beautifully illuminated the logic of evolutionary law. As she never tired of explaining, because Madagascar split from India eighty-eight million years ago, it became a distinct biosphere in which 90 percent of species are endemic, where lemurs filled all the niches that monkeys did elsewhere, and in an astonishing variety of shapes and sizes, from the dog-sized indri (which sings mournfully in tall trees) to the mouse lemur, the world's smallest mammal, to the otherworldly aye-aye, with its outsized skeletal third finger. Madagascar also hosts housecat-sized chameleons and tiny tenrec 'hedgehogs' which communicate through rubbing their striped quills together, comet orchids and the giant exploding palm tree.

David Attenborough's special interest in Madagascar has brought this wonder to the general public, though few realize that the island is nearly three times the size of Britain, and the home too of a blended Indonesian, African, and Arab culture that dates back well over a thousand years; of undulating valiha harp music in *lova-tsofina* 12/16 rhythm, of mineral mining as well as vanilla and ylang ylang farming. Antananarivo, the capital, is filled with French-inspired ice-cream-colored houses alongside traditional red clay villas, its steep streets threaded with stepped lanes overhung with bougainvillea.[5] The south boasts a warrior culture recorded first in print in the eighteenth-century bestseller *Madagascar, or Robert Drury's Journal during Fifteen Years' Captivity on That Island*, probably ghost-written by Daniel Defoe.[6] Mum loved all this, and would invariably enjoy a few days in the capital at her favourite Hotel Colbert, catching up on the gossip, before beginning the bumpy ride to her research sites.

5. E.D. Ralaimihoatra, *Histoire de Madagascar* (1966); J.-L.V. Raharimanana and C. Ravoajanahary, *Madagascar* (1947).

6. Jolly, *Lords and Lemurs*, pp. 80–81.

Despite her self-direction, my mother did not call herself a feminist, the subject of heated discussion with my sister and me for years, in which sociobiology became a line.[7] Mum's world was full of females and males doing what they were programmed to do, mine full of radical bodies reshaping relationships, and we seemed poles apart. But as she opened up to my values, I came to see the scale of her achievements, and the subtlety of her argument that biology's force is interesting precisely because it is adaptive. Ironically too, I have come to see how strongly women-centred she was. Although her tastes were for outdoor adventure, dropping Jane Austen any day for Kipling, she saw her chosen method of watching animals *in situ* as deliberately non-invasive, letting the animals live and behave 'as they wanted to.'[8] Linda Fedigan describes this as 'feminist science,' where nature is seen not as passive and subject to human control, but 'active, complex and holistic,' and in which too, the observer reflexively declares their own position.[9] It was a deliberate contrast to the laboratory-centred style of replicable experimentation in the late 1950s and early 1960s when Mum was trained. Photographs of her at Yale show her in this mould: long hair swept up in an attempt at her mother's preferred French twist, as she reaches an animal out of the cage. In fact, they never were specimens or subjects to her, and she would tell of the kinkajoo's jealousy of the lemur, until one day it escaped and ran up onto the lab window sill and was only coaxed down with a carrot. When clearing out her things, we came across four photo albums entirely of close-up portraits of ring-tails in Berenty, all carefully labelled with their names. They had been used for team observation of troops. But the fact is, she knew who

7. A. Jolly and M. Jolly, 'A view from the other end of the telescope' (1990). See also A. Jolly, 'Female biology and women biologists' (1991), pp. 39–40.
8. *Alison Jolly: Seven Wonders of the World* (1995), dir. C. Sykes, BBC.
9. L.M. Fedigan, 'Is primatology a feminist science?' (1997), p. 67.

the individuals were, and she cared about them as part of her view that, as she wrote in an undergraduate essay on plankton, 'a community of living things has a structure, even if the specific structures are too complex for him to understand. ... The student confines himself to the bounded region of a pond or log, but finds that everything affects it, from the path of a wandering newt to the climate of North America.'[10]

Despite contemporary suspicions of anthropomorphism, her approach anticipated a paradigm shift in theories of evolution. When she published her breakthrough *Science* paper in 1966, aged just 29 with me a baby on her knee, the dominant thought was that intelligence had evolved to master simple tools. She speculated that more likely it evolved through the challenge of maintaining complex social relationships.[11] As she put it in her best-selling textbook *The Evolution of Primate Behavior*, 'learning about the environment from within society and learning about society itself are thus the primate way of life.'[12] Mum certainly enjoyed using her children as subjects in this regard, and several of our 'behaviors' were analyzed as particular forms of learning, aggression, play or friendship. Looking again at this text, I see however her own behavior as a mother illuminated, following her belief that attachment is more important than food for a primate's development. Though she did pop off to Madagascar whenever she could, and always preferred an intellectual conversation to chit-chat, she refused the distant and controlled form of mothering common to her class in the 1930s and 1940s. Unsurprisingly, she couldn't always keep the balls in the air, especially when we moved to New York in 1982, and she needed to support my father's all-consuming work with the United Nations.

---

10. A. Jolly, term paper for Ithaca Habitats for Lamont Cole's ecology course, 1956.
11. A. Jolly, 'Lemur social behavior and primate intelligence' (1966). See also S. Hrdy and P. Wright, 'Alison Jolly: A supremely social intelligence (1937–2014)' (2014).
12. A. Jolly, *The Evolution of Primate Behavior* (1972), p. 355.

Her diary of 1983 records her New Year's resolutions: 'To say I like New York, when asked. To be a great hostess. To be honest, when talking, particularly about life being romantic and reality melodrama.' And a decade later, they included this reflection: 'I brought up the kids as I wanted to be—with freedom and excitement and little nagging about rules and appearance. Clearly this was wrong. But how wrong? Should I not have done what I did about career? Or tried to make NYC more of a home? Or what?' Her resolutions for 1992 included: 'Not to moan about the empty nest.'

Donna Haraway's interesting analysis of my mother's place in the history of primatology locates her achievements in the privileges of class and race, her heterosexually conventional marriage and dad's financial support, as well as the example of her own mother. Yet Haraway also describes how Mum's 'hybrid arrangement' depended on the 'craft production' that primatology still permitted in the 1960s and 1970s, in contrast to the more industrialized sciences, and that Mum's happiness with the arrangement was undoubtedly in part because of her strong self-esteem. She also pays tribute to my mother's growing self-awareness, her 'polyphonic' writing, responding to decolonization as well as gender politics by including the voices of Malagasy colleagues as well as those of the animals.[13] My mother described travelling with the botanist Rachel Rabesandratana, both leaving their children to their husbands' care, discovering similar privileges as university-educated urban women juggling kids and career. Was camping deliciously romantic (Mum)? Or dangerous and uncomfortable (Rachel)? And what did the village grandmother serving them rice think? Haraway considered that my mother, had 'taken a modest, concrete part' in the development of a properly local

13. D.J. Haraway, *Primate Visions: Gender, Race, and Nature in the World of Modern Science* (1989), p. 272.

and global conservation about its many competing lives and peoples.[14]

So the question of mothering was not the biggest, important though it was. As proof that women do not go into primatology because they like 'big brown eyes,' my mother's real interest was to discover ways that competitive species coexist and indeed cooperate.[15] She proposed that 'as we cannot reverse evolution, we have no choice but to continue using our knowledge to accept the responsibilities for our society.'[16] In *Lucy's Legacy*, which she wrote in her office at Princeton—where she had installed her mother's green chaise longue—she argued that to learn where we go next, we need to understand and harness the evolution of sex and intelligence, cooperation and love.[17] It is thus unsurprising that she moved from lemur observation to conservation. She worked with photographers, notably Frans Lanting,[18] and television crews, but also listened to and interviewed local Antandroy people, the 'lords' of Southern Madagascar, as well as the de Heaulmes, the French–Malagasy family who own the Berenty Reserve and who had become firm friends. She did oral histories of guardians, native guides, plantation workers, villagers, fishermen, farmers, anthropologists, conservationists, ecotourists, aid workers, ministers and traders and developers.[19] She lobbied. Most of all, she threw herself into training programmes with and for Malagasy biologists and teachers. Her last project was the *Ako* series (2005–12), illustrated children's books of lemur adventures in the

14. See also J. Scardina, *Wildlife Heroes: 40 Leading Conservationists and the Animals They are Committed to Saving*, Philadelphia, PA, Running Press 2012, and A. Labastille, 'Eight women in the wild,' *International Wildlife* 13, 1983 36–43.

15. On the 'big brown eyes' myth, see L.M. Fedigan, 'Science and the successful female: why there are so many women primatologists,' *American Anthropologist* 96, 1994 529–40.

16. Jolly, *The Evolution of Primate Behavior*, p. 357.

17. A. Jolly, *Lucy's Legacy: Sex and Intelligence in Human Evolution* (1999).

18. F. Lanting, A. Jolly and J. Mack, *Madagascar: A World Out of Time* (1990).

19. Jolly, *Lords and Lemurs*.

threatened forests, which she wrote with her colleague Hanta Rasamimanana, doing what they could to integrate them into teacher training programmes.[20] She had been dreaming of this since 1964, after learning that Malagasy children only had pictures of European rabbits in their books, and the only available lemur photographs in the Malagasy Republic were on the covers of match books for tourists, too expensive for locals.[21] Mum's Malagasy students were far more important to her than any of her books.

Her broadening of interest was also driven by what was happening in Madagascar. It became officially independent in 1960, two years before she first visited, but French officialdom still dominated until 1972, when, amid popular unrest, army chief Gabriel Ramanantsoa seized power as head of a provisional government. The country's ties with France were loosened in favour of the Soviet Union; French scientists were told to leave. The French government responded by taking everything with them, except a few cars and buildings.[22] The new government then denied all research visas between 1975 and 1983. Mum continued to go in on tourist visas, beginning with a six-month visit in 1975 supported by the World Wildlife Fund to write *A World Like Our Own*.[23] This was the only time she took all the family with her; four children all aged under eleven, thrilled that we would live in the country's zoo, in a whitewashed house next to Georges Randrianasolo, zoo

20. M. Jolly, A. Jolly and H. Rasamimanana, 'The story of a friendship,' *Madagascar Conservation & Development* 5 (2010), pp. 125–6. See also A. Jolly, H. Rasamimanana and D. Ross, *Ny aiay ako (Ako the Aye-Aye)*, Myakka City, FL, Lemur Conservation Foundation 2005. For more on the Ako Project, see www.lemurreserve.org/akoproject2012.html.

21. My mother in this book pays tribute to her forerunner Barthélémy Vaohita, World Wildlife Fund Representative for Madagascar in the 1980s, whose primary school readers called *Ny Voara (Nature)*, were distributed to the provinces in 1987.

22. Jolly, *Lords and Lemurs*, pp. 182–3. Notably, the French–Malagasy de Heaulme family, who were farmers not scientists, but who own the Berenty Reserve, did not leave but committed themselves to working with the new government.

23. A. Jolly, *A World Like Our Own: Man and Nature in Madagascar* (1980).

director and an important catalyst for research. My father found some work with the government on poverty alleviation. But this was also the year that Didier Ratsiraka took power after a coup which set him up for a 22–year dictatorial rule. Mum's instinctively literary rather than political sensibility was thus forced into focus by such visible nationalism, encouraged also by my father, with whom she had already begun to ask, who pays for conservation—and who benefits?

The book that follows begins with this question, with her arrival at the University of Antananarivo's 1985 conference on Environment and Sustainable Development. The Minister of Waters and Forests, Joseph Randrianasolo, explains that this conference would not be like the previous one of 1970, which had emphasized the uniqueness, beauty, and scientific interest of Malagasy flora and fauna. In contrast, he puts it that the Malagasy want to manage their resources to be self-sufficient in food and fuelwood. Realizing that this was a turning-point, Mum began to keep frank and detailed diaries.[24] The arc starts in Part I, Villages, where she reveals the influences of the outside world on apparently timeless village life. In Part II, Politics, she goes behind the scenes of the development of a National Environmental Action Plan introducing protagonists Russell Mittermeier, Tom Lovejoy, and the Napoleon-complexed Minister. The World Bank, USAID and other donors pledge funds, though not without acrimony. Then, with the funding established, Part III, Environment and Development, takes us into life at the National Park of Ranomafana in the eastern rainforest. Ranomafana's research and science are justly famous, as are its golden bamboo-lemurs. However, like all the other early international conservation and development projects, trying to bring development to unconvinced villagers is a

24. The full diaries are available as part of the Alison Jolly Papers, Cornell University Library.

fraught process. Part IV, Weather, moves south to the spiny forest—the name my mother coined—and her own research site Berenty, as well as the Bezà Mahafaly reserve championed by Alison Richard. People and lemurs alike suffer in ferocious droughts. Evolution has shaped the lemurs and the culture the people to deal with such recurrent catastrophes. Climate change raises even greater challenges in the future.

My mother concludes that traditional life is in fact unsustainable. Change will come for good or bad. Thus in the final section, Money, her diaries track the prospect of mining concessions, as Madagascar's extraordinary natural mineral resources entice the interest of outside investors. Here we also see how Mum came to be controversially involved with Rio Tinto in the development of the QMM titanium mine on the country's southern coast, as an advisor on the independent Biodiversity Committee.[25] She talked to us about this as her chance to grasp the political opportunity—even as, by then, she was in her seventies and increasingly unwell. The Committee's achievement, in her view enormously due to the skill of fellow advisor Léon Rajaobelina, a former ambassador and Minister of Finance, was to negotiate the company's commitment to net positive improvement in both environment and society during the life of the mine and a pledge not to cause the extinction of any species—an astonishing ambition. It seems this has been achieved, but Mum knew that the story might end in tragedy. She died nine days after the election of a new president, Hery Rajaonarimampianina, on 26 January 2014, hoping not.

The reader will see that these are diaries of witness, not confession.[26] Even as she recorded the dinners and the dreams, what was most important to Mum was the ecology

25. Bezanson, Keith et al., 'Report of the International Advisory Panel on QMM' (2012); R. Harbinson, 'Development recast' (2007).
26. Jolly, 'The narrator's stance' (2011).

of community, not the private life. This was evident even in her last days, during which she was frantically writing the last pages on her deathbed, installed in her beloved study. Once, she woke to tell us she had been dreaming of indri lemurs singing in the trees by Cayuga Lake in Ithaca, though she knew this could not be possible. She left us a memorial service script in which she asked that we play a recording of those indri, followed by 'If I had a hammer, I'd hammer out justice' and a recessional of cheerful trumpet music. Later we heard from colleagues in Berenty that on the day of her death, the lemurs had come out of the forest, behaving oddly. Hanta said tartly this was 'Disney.' Mum would have too—she was adamantly against fantasies of an afterlife. She didn't need one. Reality was fantastical enough and evolution will take care of the rest.

I hope that readers will see this view as her legacy, and do what they can to continue to visit, live in, love, and care for Berenty and Madagascar as a whole, as well as to marvel at its lemurs, whose fate depends upon us 'cruel but wonderfully amusing' humans. This was certainly her fervent hope, as she recorded her thoughts in these diaries through the tiny flash of a human lifetime.

Margaretta Jolly

Chronology of events

160 million years ago: Madagascar separates from the African mainland.

80 million years ago: Madagascar breaks away from India.

4,000 years ago: Madagascar settled by people of Indonesian/African descent.

800	Arab merchants begin trading along the northern coast.
1200	Central highlands are settled.
1500	Portuguese captain Diogo Dias is first European to land on Madagascar, blown off-course on the way to India. He names the island St Lawrence.
1500s	Portuguese, French, Dutch, and English attempt to establish trading settlements; they fail due to hostile conditions and fierce local Malagasy.
1880s	France consolidates its hold over Madagascar in the face of local resistance.
1910s	Growth of nationalism; discontent over French rule.
1927	Ten *Réserves naturelle intégrales* covering 160,580 ha created under French colonial order. Scientific entry permitted but local use prohibited.
1936	Forest reserves of Berenty Estate established by the de Heaulme family in consultation with Tandroy clans; sisal plantation founded beside the Mandrare river.
1937	*Alison Jolly born in Ithaca, New York.*
1946	Madagascar becomes an Overseas Territory of France.
1947	French suppress armed rebellion; thousands are killed.
1956	*Réserves speciales* creates new protected areas.
1958	*Parcs nationaux* created.

Madagascar votes for autonomy.

Alison Jolly graduates with B.A. in Zoology from Cornell University.

1960 Independence, with Philibert Tsiranana as Madagascar's first president.

1962 *Alison Jolly graduates with a Ph.D. in Zoology from Yale. First visits Madagascar for postdoctoral research on ring-tailed lemurs, New York Zoological Society.*

1963 *Alison marries English economist Richard Jolly. Four children born 1965, 1967, 1969, 1971.*

1966 *Jolly publishes 'Lemur Behavior: A Madagascar Field Study.'*

1970 'Malagasy Nature, World Heritage' Conference, University of Antananarivo, which Jolly attends.

1971 *Jolly becomes research associate, University of Sussex.*

1972 *Jolly publishes 'The Evolution of Primate Behavior.'*

Popular unrest in Madagascar. Tsiranana dissolves government; General Gabriel Ramanantsoa becomes head of provisional government. He reduces the country's ties with France in favour of the USSR.

June: Madagascar participates in the United Nations Stockholm conference on the Environment.

1975 Lieutenant-Commander Didier Ratsiraka is named head of state after a coup, and elected president for a seven-year term. The country is renamed Democratic Republic of Madagascar.

Jolly takes the family to Madagascar for six months to research 'A World Like Our Own.'

1976 Ratsiraka forms the Arema Party. He nationalizes large parts of the economy, until 1986, when market economy promoted.

1980 *Jolly publishes 'A World Like Our Own.'*

First International Monetary Fund bailout.

1982 *Jolly family moves to New York as Richard becomes deputy director of UNICEF. Alison becomes guest investigator, Rockefeller University.*

Jolly travels in Madagascar with the BBC to work on 'Tropical Time Machine' for Horizon (1983).

1984 In Madagascar, *Stratégie Nationale pour la Conservation et le Développement Durable* (SNCD) adopted.

1985 International Conference on Conservation and Sustainable

Development, Madagascar, attended by Prince Philip, Duke of Edinburgh, as president of WWF.

Jolly awarded Chevalier de l'Ordre National de Madagascar.

1987 *Jolly is visiting lecturer, Princeton University, 1987–2000.*

1989 Madagascar develops National Environmental Action Plan (NEAP).

1990 Paris: the world's first NEAP Accord signed for $100 million, ratified as *Accord pour l'Environnement*. Beginning of a period integrating conservation and development projects.

1991 Ranomafana National Park formed. ANGAP (*Association Nationale pour la Gestion des Aires Protegées*) set up to manage Madagascar's protected areas system.

President Ratsiraka forced to give up powers after army opens fire on demonstration.

1992 *Plan d'Action Environnementale* (PAE): Madagascar aims to develop a biodiversity offset policy for mining and logging companies along with other environmental incentives.

Constitution of the Republic of Madagascar is passed. Article 39 states: 'Everyone shall have the duty to respect the environment; the State shall ensure its protection.'

Madagascar participates in the United Nations conference on the Environment, Rio de Janeiro.

Under pressure, Ratsiraka introduces democratic reforms.

Alison Jolly becomes president of the International Primatological Society.

1993 Albert Zafy elected president of Madagascar.

1994 Ministry of Environment established. MECIE law (*Mise en Compatibilité des Investissements avec l'Environnement*) set up to protect the environment during development.

1996 *Jolly becomes Honorary Chairman of the International Committee, 27th IPS Congress, Madagascar.*

GELOSE law approved, seeking to integrate rural people into forest management.

1997 Zafy impeached. Ratsiraka voted back into office.

1998 International Primatological Congress, University of Antananarivo.

Jolly awarded Officier de l'Ordre National de Madagascar.

1999 *Jolly joins independent advisory panel to QMM, Quebec Madagascar Minerals, a mining operation jointly owned by Rio Tinto and the Government of Madagascar.*

Jolly publishes 'Lucy's Legacy: Sex and Intelligence in Human Evolution.'

2000 *Richard and Alison Jolly return from New York to Lewes, Sussex. Jolly becomes Visiting Senior Research Fellow, University of Sussex, until her death.*

Thousands homeless after two cyclones hit the island and Mozambique in March.

2001 May: Senate reopens after twenty-nine years, completing the government framework of presidency, national assembly, senate and constitutional high court provided for in the 1992 constitution.

December: First round of presidential elections. Opposition candidate Marc Ravalomanana claims an outright victory.

2002 January: Ravalomanana and his supporters mount a general strike and mass protests.

February: Ravalomanana declares himself president after weeks of political deadlock with Ratsiraka over the December polls. Violence breaks out.

April: constitutional high court declares Ravalomanana winner of the December polls after a recount.

June: the USA recognizes Ravalomanana as president.

July: Ratsiraka seeks exile in France.

December: Ravalomanana's party, I Love Madagascar (TIM), wins a parliamentary majority.

2003 February: Former head of the armed forces is charged over an attempted coup against President Ravalomanana.

September: World Parks Congress in Durban. Ravalomanana unveils a plan to more than triple the country's total protected area, from 1.7 to 6 million ha, by 2008. The 'Durban Vision' sees the start of a period of mainstreaming environmental thinking into macroeconomic planning, lasting until 2008.

2004 February–March: Tropical cyclones Elita and Gafilo hit; thousands are left homeless.

October: World Bank, International Monetary Fund say they are writing off nearly half of Madagascar's debt.

Jolly publishes 'Lords and Lemurs: Mad Scientists, Kings with Spears, and the Survival of Diversity in Madagascar.'

2005 *Jolly with Hanta Rasamimanana publishes 'Ako the Aye-Aye.'*

Madagascar is the first state to receive development aid from the USA under a scheme to reward nations considered to be promoting democracy and market reforms.

With contributions from the Malagasy government, Conservation International and the WWF, the private Madagascar Biodiversity Fund is founded.

August: QMM mine project gets the go-ahead from Rio Tinto.

December: Malagasy Minister of Environment, Water, and Forests creates three new protected areas, bringing a further 875,000 hectares under protection.

2006 *Microcebus jollyae, Jolly's mouselemur, named by Edward Louis.*

January: Madagascar introduces a new park-management system, the *Système d'Aires Protégées de Madagascar* (SAPM), to replace ANGAP.

June: Conservation International Conference, Antananarivo. Ravolomanananana commits to further protection of national parks.

December: Officials declare Ravalomanana winner of presidential elections.

2007 April: referendum endorses increase in presidential powers.

July: President Ravalomanana dissolves parliament after new constitution calls for end to autonomy of provinces.

September: Ravalomanana's TIM party wins 106 seats out of 127 in early parliamentary elections.

November: President Ravalomanana opens $3.3 billion nickel cobalt mining project in Tamatave, said to be largest of its kind in the world.

Jolly acts as scientific advisor on 'Lemur Street' series.

2008 February–March: Cyclone Ivan kills 93 and leaves 332,391 homeless. UN launches $36 million appeal for affected areas.

March: Madagascar produces first barrels of crude oil in sixty years.

2009 January: Dozens killed in protests in the capital. Opposition leader Andry Rajoelina calls on the president to resign, and proclaims himself in charge of the country.

February: Dozens killed after police open fire on opposition demonstration in the capital.

March: Rajoelina assumes power with military and high court backing. Move is condemned internationally and isolates Madagascar.

June: Marc Ravalomanana, who has been living in exile, is tried and sentenced in absentia for abuse of office.

August: International mediators broker power-sharing agreement in Mozambique. Deal fails.

2010 March: African Union imposes targeted sanctions on Rajoelina and his administration.

May: Rajoelina sets a timetable for a constitutional referendum and elections.

June: EU decides to suspend development aid in the absence of democratic progress.

August: Marc Ravalomanana is sentenced in absentia to life in prison for ordering killings of opposition supporters.

November: Voters in referendum endorse new constitution that would allow Rajoelina to run for president.

2011 September: Eight political parties sign agreement to pave the way for elections to re-establish democracy. The deal leaves Rajoelina in charge of a transitional authority until March 2012 elections; it also allows for the return of the exiled Ravalomanana.

November: New unity government is unveiled. Opposition parties agree to join new government 'with reservations.'

Former president Didier Ratsiraka returns after nine years in exile.

2012 *Jolly awarded honorary doctorate from the University of Antananarivo.*

May: Andry Rajoelina says he hopes elections can take place 'as soon as possible.'

June. Madagascar attends the United Nations Conference on Sustainable Development (UNCSD), also known as Rio+20 or Earth Summit 2012, aimed at reconciling economic and environmental goals of the global community.

September: Amnesty International calls on government to rein in security forces accused of killing dozens.

2013 January: Andry Rajoelina and Marc Ravalomanana agree not to contest elections, following SADC. When Ravalomanana's wife Lalao announces her candidacy, Rajoelina announces he will stand, as does Didier Ratsiraka. All declared invalid.

August: International Prosimian Congress, Ranomafana.

2014 January: Hery Rajaonarimampianina sworn in as president after elections.

February: Alison Jolly dies at home in Lewes.

Dramatis personae

ROLAND ALBIGNAC, French academic. Director of UNESCO's Man and the Biosphere Project. Key to achieving national park status for the Mananara Reserve.

JOSEPH ANDRIAMPIANINA, forest specialist. School of Agronomy, University of Antananarivo; head of the National Office of the Environment in the 1980s.

DAVID ATTENBOROUGH, BBC wildlife presenter, pioneering broadcaster and naturalist.

JOSEF BEDO, naturalist, guide and son of chief forester at Perinet.

DENNIS DEL CASTILLO, Peruvian agronomist and conservationist; head at the Amazon Research Institute.

DE HEAULMES, French–Malagasy owners of Berenty Reserve and sisal plantation. Three generations of the de Heaulme family have preserved the Berenty Reserve since 1936, and welcomed scientists since 1963.

FRANÇOIS FALLOUX, senior environmental advisor, Africa Region at the World Bank, and architect of the National Environmental Action Plan, Madagascar.

JÖRG GANZHORN, German ecologist and conservationist associated with Tsimamampetsotsa Reserve.

LISA GAYLORD, environment and development expert who has worked for USAID, World Conservation Society and Rio Tinto QMM.

JOLLY family: Alison's husband Richard and children Margaretta, Susan, Arthur (Morris) and Richard (Dickon).

FRANS LANTING, Dutch photographer, best known for his outstanding wildlife photographs in *National Geographic*.

Tom Lovejoy, pioneer in the science and conservation of biological diversity; originator of the concept of debt-for-nature swaps. Formerly director of the World Wildlife Fund US program; American University Professor of Environmental Science and Policy.

Bernhard Meier, German biologist; co-discoverer of the golden bamboo lemur, Ranamanfana.

Russell Mittermeier, American primatologist and herpetologist. President of Conservation International.

Emma Napper, BBC Earth producer for *One Planet*, including *Madagascar* (2011).

Toshisada Nishida, Japanese primatologist. Head of Evolution Studies at Kyoto University in Japan; former president of the International Primatological Society.

Joe and Dai Peters, American conservation resource managers and consultants in conservation education and development. Worked at Ranomafana National Park.

Jean-Jacques and Arlette Petter, French primatologists who pioneered the study of lemurs.

Léon Rajaobelina, Malagasy economist and conservationist. Formerly governor of the Central Bank, minister of finance during the elaboration of the NEAP, ambassador to the USA in the 1980s, and vice president of Conservation International.

Émile Rajeriarson, Malagasy researcher and guide at Ranamafana National Park; co-discoverer of the golden bamboo lemur with Bernhard Meier. An amphibian is named after him.

Ny Fanja Rakotomalala, Malagasy engineer. President of Rio Tinto QMM.

Berthe Rakotosamimanana, Malagasy primatologist. Permanent Secretary of Higher Education, Secretary General of the Groupe d'Études et de Recherche sur les Primates de Madagascar (GERP) from its founding until her death in 2005.

Johny Rabenantoandro, Malagasy botanist. Environment manager of Missouri Botanical Gardens, later Rio Tinto QMM.

Jean-Baptiste Ramanamanjato, Malagasy herpetologist; biodiversity and rehabilitation superintendant, Rio Tinto QMM.

Guy Ramanantsoa, Malagasy zoologist and herpetologist. Chief engineer dealing with water resources and national parks (1970). Chair of the Water and Forestry Department in ESSA, the agronomy school of the University of Antananarivo.

Georges Randrianasolo, Malagasy ornithologist. Director of the Parc Tsimbazaza, Antantanarivo national zoo.

Joseph Randrianasolo, Malagasy politician. Minister of Eaux et Forêts, 1980s.

Loret Rasabo, chief guide at Ranamanfana National Park.

Hanta Rasamimanana, Malagasy primatologist; close colleague and friend of Alison. Professor at the École Normale Supérieure, University of Antananarivo. Malagasy author and leader of the education component of the Ako Project with Alison Jolly.

Joelisoa Ratsirarson, Malagasy ecologist. Professor at the Forestry Department of the School of Agronomy at the University of Antananarivo, and vice president of the University.

Gilbert Ravelojaona, associated with Bezà Mahafaly Reserve. President of ESSA, School of Agronomy, University of Antananarivo.

Alison Richard, British primatologist; founder and lead scientist of Bezà Mahafaly Reserve. Vice-Chancellor emerita of Cambridge University; senior research scientist, Yale University.

Eleanor Sterling, chief conservation scientist, Center for Biodiversity and Conservation, American Museum of Natural History.

Philibert Tsimamandro, anthropologist of the Tandroy people in the south of Madagascar.

Patricia Wright, American primatologist, anthropologist, and conservationist best known for her study of social and family interactions of wild lemurs in Ranomafana National Park.

Barthélémy Vaohita, representative of the WWF in Madagascar in 1980s–90s; now président de l'Alliance française d'Antsiranana.

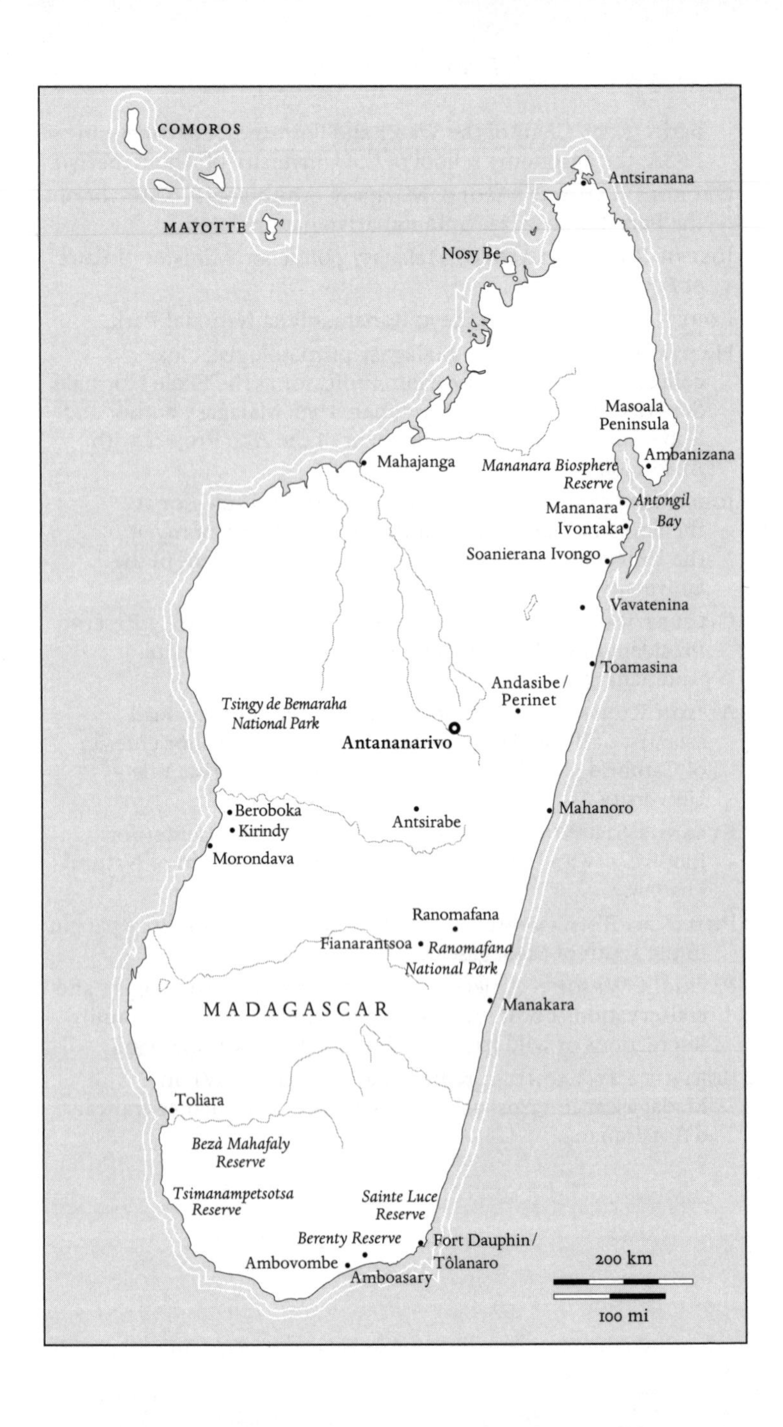

COMOROS
MAYOTTE
Antsiranana
Nosy Be
Masoala
Peninsula
Ambanizana
Mahajanga
Mananara Biosphere
Reserve
Mananara
Antongil
Bay
Ivontaka
Soanierana Ivongo
Vavatenina
Toamasina
Andasibe /
Perinet
Tsingy de Bemaraha
National Park
Antananarivo
Beroboka
Kirindy
Antsirabe
Mahanoro
Morondava
Ranomafana
Fianarantsoa
Ranomafana
National Park
Manakara
MADAGASCAR
Toliara
Bezà Mahafaly
Reserve
Tsimanampetsotsa
Reserve
Sainte Luce
Reserve
Berenty Reserve
Fort Dauphin /
Tôlanaro
Ambovombe
Amboasary
200 km
100 mi

ONE

'Our country is committing suicide'

In Madagascar, wide-eyed lemurs offer insights into humanity's deep past. Swivel-eyed chameleons flag their emotions in scarlet and emerald and indigo blue. Lemon-yellow luna moths escape their predators by falling from the treetops like drifting leaves. A spider spins golden silk which can be woven to create a cape worthy of a Malagasy queen. In the forests of this island-continent, almost 90 percent of species are endemic, unique to Madagascar. It is an alternate world of evolution.

However, if its forest destruction continues, we could be left with only lemurs confined to zoos and the rarest palms and baobabs in botanic gardens and seed banks. The Malagasy golden-silk spider is endemic of course—not rare, but only a handful of humans know how to spin its silk. Conservationists like myself believe that the living ecosystems of Madagascar are a world heritage in the truest sense: a treasury of beauty, science and wonder for the entire world—and that keeping these ecosystems and their species alive is a responsibility for the world.[1]

At the heart of this tale is the conflict between three different views of nature. Is the treasure-house of Madagascar,

1. www.vam.ac.uk/content/articles/g/golden-spider-silk.

with its beauty and evolved fascination, really a heritage of the world? But what about the people who live beside those forests, and depend on their bounty? Is it instead a legacy of those people's ancestors, bequeathed as a sacred legacy to serve the needs of their living descendants? Or else is it an economic resource to pillage for short-term gain or to preserve in the longer term only for its return in 'environmental services'? The three attitudes—aesthetic, traditional and economic—collide throughout this book.

Thank you, Madagascar is an eyewitness account of a major case study in the politics of conservation. I argue that in spite of the irritations or even fury of the people on all sides of the story, Madagascar must save its heritage. In spite of being the world's tenth poorest country, in spite of having at the moment a government which does not govern, in spite of being a political football dribbled in any direction by foreigners offering cash or in pursuit of cash, Madagascar is one of the glorious places. Not just the forested semi-natural environment, but its cultures.

Madagascar is a blend of African, Indonesian and Arab backgrounds that dates back well over a thousand years.[2] The Indo-Polynesian tongue is studded with Bantu words, different in the differing dialects. Music ranges from plaintive polyphonic strings to African rhythms of drums and stamping heels. As in most countries, there are splits which translate to differences between people at the center and the periphery, darker people and lighter people, rural people

2. A note on Malagasy names. Most names are a combination of words. Plateau people's names begin *Ra*, Sir, Mr, Mrs, or *Andriana*, Lord. They go on to combine stock phrases: *solo*, son; *koto*, grandchild; *manana*, riches or possessions, etc. Thus, *Rasamimanana* means Mrs We-Share-Riches. Coastal names may omit the prefix and may be more inventive. Translated to English they are just as long as in Malagasy. Similarly place names are composites. *Antananarivo*, the capital, is *an*, place of; *tanan*, city; *rivo*, a thousand—or Very-Big-City. The large soda lake *Tsimanampetsotsa* is *tsi*, not; *manam*, has; *petsotsa*, whales. Vowels are pronounced much as in French, except for *o*, which is said *oo*. Stress is usually on the antepenult, so the end of the word disappears to European ears.

and those in the towns. Malagasy are justly proud of their complicated nation—but not all know that the nation needs its environment to survive. Sustainable development for both people and natural environment is the only worthwhile goal. Madagascar may fail, sinking ever deeper into poverty and degradation, but the stakes are so high that it ought, somehow, to win.

The book is not a polemic. It is a story. In 1985 I watched a breakthrough: a moment when national politics and foreign funding came together. Now I knew I was eyewitness to conservation history. So I wrote it down. I scribbled in airplanes, on hotel tables, in tents under the mosquito netting. I have drastically cut and excerpted and added a few explanatory phrases, but the diaries are what I saw and felt at the time. (The original notebooks, twice as long, are stored in the Cornell University Library archive.) If my style seems a bit too literary, actually I always write like this—at least ever since my long letters home to my parents at the age of 25 when I first discovered Madagascar as a magical world. The adventure tales of my childhood paled beside adventure I could live: with people, with landscape, with lemurs![3]

I thought these diaries couldn't be published for twenty years or so until my friends were old and gray and would not care what I said about them. Twenty years have more than passed. My friends are old and gray and feistier than ever.

3. My father, Morris Bishop, wrote biographies of Pascal, Ronsard, La Rochefoucauld, Petrarch, and the explorers Champlain and Cabeza de Vaca. Also *A History of Cornell*, the wry introductions to each author in his *Survey of French Literature*, and light verse for the New Yorker. He read aloud to me about Kipling's Kim tramping with his llama on the dusty roads of the British Raj, *Treasure Island*, and Rider Haggard's lost kingdoms of Africa. My mother, Alison Mason Kingsbury, was a painter who seemed as familiar with the whole of art history as my father was with the history of European literature. I went into biology in part to find something to talk about that they did not already know—but also from the early conviction that science was even more outrageously improbable than anything in my parents' verbal and visual worlds.

Now I think it is time to tell you my story of a time when conservationists thought we could save the world—or at least Madagascar.

❦

I arrived in Madagascar in 1962 as a stunningly ignorant Yale Ph.D. Like others of my kind I was single-minded: I wanted to watch lemurs, not people. Besides, I was in love. I hoped to get married and live happily ever after, far from Madagascar—if only Richard Jolly would make up his mind to propose! Richard was and is an economist who actually likes people.

I wrote up my fieldwork, but I was soon prevented from undertaking further field study by the arrival of our four children. I pottered away in England writing second-hand science. I fretted over my distance from Madagascar, but I didn't have a chance to return until 1970 for the first of four international conferences that anchor the story of this book: 1970, 1985, 1998, and one happening as I write in 2013. The 1970 conference was 'Malagasy Nature, World Heritage'.[4]

Jean-Jacques Petter and Monique Pariente organized that meeting. Jean-Jacques and his wife Arlette, both from the National Museum of Natural History in Paris, had done the only detailed study of lemur ecology and behavior that preceded my own. Monique Pariente was almost ridiculously well connected. Her father, General Gabriel Ramanantsoa, would become the country's president in circumstances no one foresaw in 1970.[5]

We met in the national university, poised on its pleasant hilltop. The view in one direction embraced the capital city,

4. A. Jolly, *Lemur Behavior* (1966); A. Jolly, *The Evolution of Primate Behavior* (1972).

5. J.J. Petter, 'Recherches sur l'écologie et l'éthologie des lémuriens malgaches' (1962); A. Petter-Rousseaux, 'Recherches sur la biologie de la réproduction des primates inférieurs' (1962). Since this book essentially starts in 1985, I won't even begin to acknowledge the giants of French science who pioneered the study of Malagasy nature. Jean-Jacques and Arlette must stand as representatives of a long and glorious tradition.

Antananarivo, its skyline culminating in pre-colonial royal palaces. The opposite view opened to a panorama of rolling hills clothed in dry golden grass ready for annual burning. Cupped in the valleys glowed green rice paddies. Gardens of pink hibiscus and orange marigolds softened the concrete university buildings.

France had colonized Madagascar as late as 1895. The country achieved independence in 1960. French policies lived on. The forests were declared to belong to the nation. Forestry action attempted to suppress indigenous slash-and-burn, called *tavy*, both to promote French commercial interests and to save what French scientists had long recognized as extraordinary natural heritage. Madagascar boasted some of the first nature reserves in the whole African region, established from 1927. A lovely zoo, a botanical garden and a research institute in the capital itself also dated from 1927, encircling a traditional sacred lake. In 1970 there were still many French scientists and Frenchmen behind government office doors, two hundred of them on the staff of President Tsiranana alone. The French ambassador, not the Malagasy president, lived in the ex-governor general's palace. The splendid new university was still the Université Charles de Gaulle.[6]

Richard had the idea of a joint paper for me to present at the conference 'Conservation: Who Benefits and Who Pays?' In the short term, we said, tourists and tour companies benefit from national parks, while peasants lose rights to their land. A banal thought, nowadays, but the reaction then was interesting. The stars of the conference surged up on either side of me: the solo transatlantic aviator Charles Lindberg, then president of the World Wildlife Fund, and Sir Peter Scott, the great ornithologist and artist, who drew WWF's Panda logo. They marched me out into the university gardens for

6. M. Brown, *A History of Madagascar* (1995).

discreet reproof. 'Of course we know what you say is true, but this is not the time to say it. We have only just convinced East Africa to promote their own national parks. We don't want them getting the wrong idea. Don't stir up trouble.' I was too overawed to protest.

Then, no sooner inside again than Perez Olindo, head of the Kenyan Game Department, and David Wasawo, soon to be president of the University of Dar es Salaam, loomed up on either side of me in turn. They marched me back round the gardens, out among the marigolds. 'High time somebody said that!' they exclaimed. 'Come and stay with our families in Nairobi!' Somehow someone left that paper out of the published conference proceedings.[7]

Meanwhile, trouble was brewing—particularly among the students privileged to study in the Université Charles de Gaulle. In 1972 they erupted into violence. They wanted the University nationalized. They wanted jobs on graduation, not Frenchmen standing in their way. Student mobs ruled the streets of Antananarivo. The town hall burned down. The cavity left by its blackened carcass remained on the Avenue de l'Indépendence for almost four decades. Monique Pariente's father, General Ramanantsoa, became temporary president, followed in 1975 by the hardline socialist Richard Ratsimandrava. Ratsimandrava was assassinated after nine days' rule. The state council then elected the almost equally socialist frigate captain Didier Ratsiraka.

Ratsiraka hoped to become a famously idealistic francophone Nyerere. Instead, socialism did not work in Madagascar. There were many reasons, external as much as internal. In 1982 the IMF and the World Bank took over the bankrupt

---

7. A. Jolly, 'The narrator's stance: story-telling and science at Berenty Reserve' (2011).

state. They imposed harsh 'structural adjustment': what today we would call 'austerity'. The economy slid further and further into depression.

A meeting in Jersey in 1983, hosted by Gerald and Lee Durrell, began the cumbersome process of reopening research visas. It was there that Jean-Jacques Petter suggested a second great conference. WWF International had appointed his protégé Barthélémy Vaohita as their representative. Barthélémy had a huge mandate: to see to the beginnings of on-the-ground WWF programs, right down to finding spark plugs for a donated motorboat. Anglo-Saxons doubted his abilities in practical matters. Now he had a task he could do brilliantly as perhaps the only Malagasy naturalist who would intervene politically. In 1984 he convinced every minister to sign a declaration in favor of sustainable development. The government of the time finally realized it had just one new fishhook for foreign aid. The bait? Lemurs! Chameleons! Biodiversity! Never mind that their own biodiversity was a side issue to almost every Malagasy, except as it could be harvested or eaten. Somehow the rich foreigners were obsessed with it—and would pay.[8]

So the next great conference arrived in 1985. This time it was 'Madagascar: Conservation and Sustainable Development'. Five hundred civil servants from the provinces were summoned to the capital to hear about the new policy. Donors and politicians figured far larger in the mix than scientists. The president of WWF was honorary chair: Philip, Duke of Edinburgh, husband of the British Queen. The assembly waited much more eagerly for Kim Jaycox, vice president of the World Bank.

For ten years, from 1975 to 1985, the xenophobic Ratsiraka government had kept out foreign conservationists. Now we

---

8. Repoblika Demokratika Malagasy, *Stratégie malgache pour la conservation et le développement durable* (1984).

were back with a new mandate and a hope of actual funds. It was perfect timing. The World Bank had royally messed up in the Amazon. Protesters hung bloody banners opposite its Washington buildings: The Bank Murders Rainforests! Madagascar offered itself as a virgin country, or at least one unharassed by biologists for the previous ten years. Hooray for a country where the environmentalists might this time get things right.

My diaries begin…

October 25, 1985. Free Day! (Giggle).

Madagascar wins again versus American can-do! Wonderful Madame Berthe Rakotosamimanana has organized a one-day meeting of us scientists before the main Congress. This in the Solimotel, a new hostelry downtown. I used to call Mme Berthe the Tiger of Tananarive when part of her job was refusing research visas to foreigners. Then last year she turned up at Jersey Zoo, invited by Gerry and Lee Durrell. She and Barthélémy Vaohita of WWF and Joelina Ratsirarson of Water and Forests and a bunch of us scientists all brokered a cumbersome agreement to let foreign research back in. At the end Mme Berthe's round face was wreathed in round smiles. I realize this is what she'd hoped all along.

Anyhow today she came out in her true colors as Prof. of Paleontology. Russell Mittermeier, VP of the World Wildlife Fund, was to lead off the science conference. He does not trust Madagascar's technology. He brought not just slides but his own projector in his carry-on luggage. Kerfuffle as people hunted round for extension cords, etc. At last Russ triumphantly plugged in his American machine—and BANG!—fried not only his projector but all the electrics in the Solimotel. Darkness reigned.

Result: Mme Berthe calmly postponed the talks till tomorrow. She'd allowed an extra day.

And for me a chance to write up and start a new diary. How I love French notebooks with their neat vertical lines that keep my handwriting in check, not just horizontals!

Off soon to meet the gang of happy naturalists in the bar of the Hotel Colbert. But all next week I am going to be stuck in the Ministry of Foreign Affairs listening to speeches while Russ and Jean-Jacques Petter get to go gallivanting with the Duke of Edinburgh in his royal jet.

October 30, 1985. 'Your country is committing suicide!'

Prince Philip, Duke of Edinburgh, husband of Queen Elizabeth, is here as president of the World Wildlife Fund. He flew the minister of Water and Forests, Russ and Jean-Jacques westward among with Monsieur de Heaulme, who owns a plantation with an airstrip north of Morondava, and then down to Berenty.

Russ reports: Below them, grasslands and forests in flames. This year there is political unrest so there are more fires than ever. It is one thing the peasants can do to annoy the government. Top members of government don't travel much outside their own constituencies, certainly not by choice.

They landed at the De Heaulme plantation of Analabe on the west coast. Analabe means the Great Forest. It is baobab forest, where the reniala, or 'Mothers of the Forest,' tower twenty meters above the woodland that clings to their skirts. The baobabs, one of seven endemic species, are shaped like Doric columns with a swelling entasis of creamy bark.

This lot were on fire.

Actually the brush below was on fire. Some farmer simply wanted to plant a crop of maize. Adult baobabs, filled with stored water, would probably survive, even if scorched to their majestic crowns. What won't survive is their seedlings. And the giant jumping rat, a cuddly long-eared mammal like a gerbil crossed with a rabbit, which only lives in this one river valley in the whole world. The flat-tailed tortoise with its honeycomb-patterned carapace only lives here, as well. After the fire the Duke and the minister could have picked up an armful of the palm-sized tortoises baked in their shells.[9]

That was when the Duke exclaimed, 'Your country is committing suicide.'

Joseph Randrianasolo, the minister, jolted from his habitual smoothness into embarrassed fury, began to say he'd hand out punishments, install guards. Monsieur de Heaulme, the forest owner, was saying 'No, no, we must have the people's cooperation and a locally based reserve.' Jean-Jacques Petter was talking hard with Russ trying to heal up old wounds and suspicions between WWF International in Geneva and WWF–US. Probably that reconciliation was also high on the Duke's agenda in his role as president of WWF. De Heaulme and the minister reached at least a temporary truce as well, after glaring at each other for the first ten minutes over lunch.

By the end of the trip, in Berenty, relations had thawed—except with Mme Aline de Heaulme. She set a tray for breakfast for the royal visitor with her best embroidered tea-cloth, her best china, croissants specially trucked two hours from the nearest bakery, and a spray of bougainvillea in a small vase. The equerry came down, took the tray—and ate it up himself. Eventually the Duke descended to the communal table set with half-used butter pats and croissant

9. *Adansonia grandidieri*. Madagascar has seven endemic species of baobabs. All of continental Africa has only one species. Giant jumping rat: *Hypogeomys antimena*. Flat-tailed tortoise: *Pyxis planicauda*.

crumbs. He cheerfully joined the conversation and salvaged some breakfast.

Surrounded by trees full of lemurs even the minister said, 'I must come back and bring my children.'

Nov. 1, 1985. Antananarivo. The 1985 Conference on Conservation and Sustainable Development.

The huge circular hall of the Ministry of Foreign Affairs, the floor filled with dignitaries' seats, while modest rural development agents from the provinces stack the tiered balconies. Vice President Jaycox of the World Bank appeared for just 45 minutes. He made a speech. Then he jetted away. I know it's unreasonable of me to expect the Bank to listen to anyone else instead of just laying down Bank conditionality for any loan to support the environment, which was all his speech. Still, it's damned irritating.

Hundreds of civil servants are bidden to attend three days of non-stop plenary sessions of non-stop oratory, outlining the new government policy of sustainability. I sat through it taking notes for my future article in the *National Geographic*. Delegations of diplomats attended the formal opening and the closing. The conference was well organized in that it ran at all, but not so divorced from Madagascar that sessions finished on time. Not even the one before the Duke's final speech. The British ambassador squirmed in agony when the Duke came in and took his place while the hall still had people drifting out of the last session. Protocol says that everyone else waits for royalty, not the other way around. At last, though, we got a thousand of us settled, while the royal WWF chief waited cheerfully.

His opening, 'I hesitate to say...' actually did sound like hesitation, as though his feelings and reactions were so strong that he had to tell the truth about his shock and dismay, in spite of the feelings of his hosts.

'I hesitate to say how astonished I am that an island which depends so much on its forests should be so denuded, and which depends so much on water should let its waters be so polluted by erosion... All the great civilizations of antiquity succumbed for three reasons: population growth, over-exploitation of forests, followed by erosion, and over-exploitation of soils, followed by loss of fertility. Many of these ancient regions have never fully recovered... All governments must sometimes make unpopular decisions, but conservation is an integral part of development, not its alternative. Government must control exploitation, the population must have access to education and understanding... and pyromania must stop! It would be easy to despair after what I have seen, but conservationists share one trait: an almost unlimited optimism!'

There was a pause as the audience teetered, wondering whether to feel anger and embarrassment that a visitor should talk straight, or else whether to side with him. Finally Remy Tiandrazana, a member of the President's Revolutionary Council, took the microphone to say on his own account those words from the airplane, 'We are heading straight for suicide!'

The smooth, smooth minister of Water and Forests, Joseph Randrianasolo, said, 'No, let us not end this conference in bitterness, because we think of our children and we will leave them a secure future.' The Mephistophelian-visaged secretary of foreign affairs, Maurice Ramarozaka, declared, 'It is not too late, because we are here at this conference deciding to take action.'

But of course that was the end of the conference... nothing left but for somebody to take action, a rather new idea for Madagascar.

In the midst of all this, to my surprise, I was informed that I am now a Knight of the National Order of

Madagascar. I have been one of the few Western scientists who kept coming and going to Madagascar in the xenophobic years. My 1980 book is all about travels with Malagasy scientists. Is this really enough to make me a knight?[10]

There was one more reception, at the British ambassador's. Chit-chat with Gerald Durrell about receiving his own honor, royally pinned on by Prince Charles, the Prince of Wales, in the grand ballroom at Buckingham Palace. The Prince confessed to enjoying Gerry's books when he was at school, which made Gerry feel about a hundred years old.

In contrast, I went down this morning to a little hole in the wall which I thought was a sweet shop. I walked past it a couple of times before being told it was medals for the Order, not boiled candies, in the cases. I then bought myself a bar-pin of the Ordre National for wearing on an army uniform. The only remaining actual business suit ribbon had a toothed and very rusty metal backing for clipping into one's lapel, but it had come unglued, so the shopkeeper promised me another for Monday. I wonder if British knights have to buy their own ribbons? Anyway, I'm now decorated like Gerry.

And at last there was a moment to ask the Duke of Edinburgh to autograph my book in a series for which he'd contributed the foreword.[11] The British ambassador nearly had apoplexy. Royals don't give autographs—but the Duke glanced at the foreword, remarked 'I've signed that page already,' flipped back to the flyleaf and wrote 'Philip' in bold vertical strokes.

So I had a chance to say that he had done more for conservation here in 24 hours than some of us in 24 years.

10. A. Jolly, *A World Like Our Own: Man and Nature in Madagascar* (1980).
11. A. Jolly, P. Oberlé et al., eds, *Madagascar: Key Environments* (1984).

PART I

Villages

In Madagascar, some 70 percent of people still live in the countryside, not in towns. They depend on what they can grow or gather. In spite of their apparently timeless isolation, in fact every village is impacted by the wider world, including government policies that reflect the views of the elite. The elite are mainly urbanites of the Merina tribe who live on the high plateau around Antananarivo. Their own culture is based on settled rice cultivation. Paddy rice around the world goes with an involuted society: people marry cousins and neighbors. One description of a good marriage is 'sewing together the rice-fields.' For people on the coastal periphery there is traditionally a hope of new land from the forest. To the Merina (though they know it is politically incorrect to say so) that means coastal tribes can afford to be lazier and have looser morals. Meanwhile the foreign elite are steeped in our own Western attitudes.[1]

Too often those attitudes have demonized subsistence farmers who practice slash-and-burn *tavy* in the forest. The Duke of Edinburgh's cry, 'Your country is committing suicide,' can far too easily translate into 'Your shifting, shiftless poor are murdering their land. Educate them! Enlighten

1. C. Geertz, *Agricultural Involution: The Process of Ecological Change in Indonesia* (1963).

them! Show them that real progress is in settled agriculture, like us in the west, or like us with our permanent rice fields on the plateau!'

The most influential early demonizer was Perrier de la Bathie. His extensive study of Malagasy plants showed him that while forests held up to 95 percent endemic species, the grasslands of the plateau were almost all pan-tropical forms. He concluded that the grasslands are wholly man-made. His statement that some 80 percent of Madagascar was already deforested by people's carelessness is still sometimes quoted today. Modern pollen studies, instead, show that grassy tree savannah and forest cover on the plateau have always co-existed, shifting in location as climate changed.[2]

Esther Boserup summed up the fallacy of blaming shifting cultivators way back in 1965. She pointed out that this is an eminently sensible use of time and resources wherever population is low and forest abundant. Forest fallow clears weeds and pests and re-fertilizes the soil with far less effort than sweating humans and draft animals. In France in the fourteenth century plague and wars depopulated the countryside. The forest returned; wolves ranged to the gates of Paris. Then, even in haughty France, shifting cultivation made a comeback. In the northern Scandinavia of Boserup's own youth, shifting hill farmers and settled valley-bottom farmers coexisted for centuries, perfectly aware of each other's methods and intermarrying their sons and daughters. Boserup concluded that it is not education, but population growth, which forces the changes. She saw population growth as leading to intensive agriculture and then on to 'civilization,' both in the sense of clustered city living and in progress to

2. H. Perrier de la Bathie, *La végétation malgache* (1921); D.A. Burney, 'Climate change and fire ecology as factors in the Quaternary biogeography of Madagascar' (1996); D.A. Burney, 'Theories and facts regarding Holocene environmental change before and after human colonization' (1997); D.A. Burney, 'Madagascar's prehistoric ecosystems' (2003).

wider human horizons. She never claimed the transitions would be easy: it is hungry or desperate or ambitious people who change their old ways.[3]

Disapproval if not demonization underlies efforts to check Malagasy *tavy* cultivation. This began as the Merina kingdom of the plateau spread over other conquered tribes. It continued through colonial attempts to ban both forest *tavy* and plateau grass burning, and is re-created by conservationists hoping to save natural ecosystems. Recently, though, many anthropologists have taken up Boserup's theme. Burning the grassland of the plateau is a rational strategy to provide cattle with a 'green bite' of re-sprouting grass at the end of the dry season, and to clear the thorny scrub that precedes any re-sprouting of trees. In the forest, *tavy* farming has hundreds of years of tradition behind it. Just as in Boserup's Scandinavia, forest-edge farmers are well aware of the alternatives. They commonly farm both paddy rice and *tavy* fields, shifting their own efforts according to their own possibilities. Almost anywhere in the thousand miles of forest by Madagascar's east coast, people will, if asked, take you to see the sites of ancient tombs or even villages now overgrown by what looks like pristine rainforest.[4]

Enforcement of the bans on *tavy* has fluctuated with Madagascar's changing politics. Periods of draconian control meant that new forest clearing became criminal without a forester's legal authorization. Laxer periods allowed fairly free access. Of course there is a wide gray area where the forester just asks a small or large bribe for his authorization. When I travelled in 1975 to research an earlier book, one friend described the rural forester as 'the most hated man in Madagascar.'[5]

3. E. Boserup *The Conditions of Agricultural Growth* (1965).

4. C.A. Kull, 'The evolution of conservation efforts in Madagascar' (1996); C. Kull, *Isle of Fire: The Political Ecology of Landscape Burning in Madagascar* (2004); J. Pollini, 'Slash-and-burn cultivation and deforestation in the Malagasy rainforests: representations and realities' (2007).

5. Some summaries of past forestry policies are in G.M. Sodikoff, *Forest and Labor in Madagascar: From Colonial Concession to Global Biosphere* (2012); Jolly, *A World Like Our*

Chapter 2 begins my 1985 diary of travelling with an old-time forester who attempted to enforce the law on that 'demonic farmer': a man typically from the poorest family of the village. We are in the lush sea-level rainforest of the east coast Masoala Peninsula. There the pockmarks of forest *tavy* have spread on many slopes until the forest is gone, although the area has long been classed as a reserve and is now a national park.

In Chapter 3, also 1985, I went west to the village where the Duke of Edinburgh watched baobabs burn. It traces some of the interplay between a semi-abandoned European plantation and the farmers' recourse to forest clearing, but mostly it records a fearful season when children needlessly died. The diary of Chapter 4 returns in 2010 to that same village, now focused on farming the ex-plantation fields. A BBC team with David Attenborough filmed Madame Berthe's pygmy mouse lemur, arguably the world's cutest mammal. I was by then in my seventies and David in his eighties—just as excited as ever when a charismatic schoolteacher read to her eager class the little book I wrote about Berthe's mouse lemur.

Chapter 5 switches back again to 1987, to the east coast of the island. A village can indeed seem to visitors like a romantic dream. In Ivontaka people tolerated wild aye-ayes among their coconut palms—even as the first plans for a new Biosphere Reserve were taking shape. I am a biologist, though. I still thought mainly about lemurs, not people—which those who watch the Attenborough programs still do. People were, and to the animal-doting north often are, just context.

Own; Kull, 'The evolution of conservation efforts in Madagascar'; Kull, *Isle of Fire*; J. Pollini, 'The difficult reconciliation of conservation and development objectives: the case of the Malagasy Environmental Action Plan' (2011); S.J.T.M. Evers, C. Campbell et al., 'Land competition and human-environment relations in Madagascar' (2013).

TWO

Dancing in the rainforest

In 1985 photographer Frans Lanting and I set out to write a story for the *National Geographic*. Unusually for that time we weren't just describing natural wonders. We intended to write about conservation.[6]

We started with the tallest rainforest at the lowest altitude, where the Masoala Peninsula plunges down near-vertical slopes to the Bay of Antongil. The slope goes straight on down into the deep water of the bay: here humpbacked whales migrate to bear their calves, sheltering behind the peninsula from the storms of the Indian Ocean. The forest above echoes with the roaring of red ruffed lemurs, improbably furry creatures the size of overfed tomcats. Wild red ruffs are almost wholly confined to the Masoala, although you may see them in zoos. Perhaps you have already cooed over their soulful yellow eyes and their bushy red fur, or ducked as they suddenly bellowed as loud as lions.[7]

We'll return in the final chapters to the Masoala's dreadful fate, but in 1985 the loggers and farmers were ordinary people, nibbling at the forest bit by bit for their own small ambitions. We travelled with a forester still trying to enforce legal

6. A. Jolly, 'Madagascar, a world apart (1987).
7. Red ruffed lemur: *Varecia rubra* or *Varecia varecia rubra*.

authority over land long designated as a National Reserve, in order to meet just one farmer to represent here thousands of others like himself—poor, hopeful, and trying to feed his family.

Friday, Sept. 27, 1985. Masoala Peninsula. The illegal clearing.

Martial Ridy may face a fine or penal servitude for cutting trees. It isn't all because of us, but we rubbed their noses in his misdeeds.

As we chugged along in the *African Queen* (its motor says *Petraka Petraka Petraka*)[8] we saw a new bare slope just behind Ambanizana village. Brown soil where light-loving plants hadn't sprouted, and a mess of tree trunks higgledy-piggledy. David Ratiarson, our acute companion from the Maroansetra Water and Forests, had already been notified there was unauthorized clearing here. One of the reasons he was glad to come with us was to look into the matter. Of course he had no budget to hire a boat for himself.

We sailed along the roadless Masoala coast. The boat looked about as sturdy as son Arthur's last home-made go-kart: paint peeling, and the top plank on the starboard side held on with rope. The motor stank of diesel. Our captain wore an enormous straw hat suitable for an English summer wedding, which curled up in front like a cavalry hat. His eyelashes curled to match. We'd first tried stopping at Hiaraka village, where I was marooned for a week in a storm ten years ago, but the slopes above it have since been cut bare with no forest left to fell. So now we were headed for Ambanizana.[9]

Landing was simple. The captain just ran the *African Queen* onto the beach. The crew put a home-made wooden ladder over the bow and we climbed down into warm water.

8. J. Thurber, 'The Secret Life of Walter Mitty.'
9. Jolly, *A World Like Our Own*.

Formal courtesies with the village president and then pressure from him and from forester Ratiarson. Martial Ridy, young and poor, had no choice but to show us his new-cut field.

Though frightened, Martial led us up through the rainforest, armed with ax and *coupe-coupe* and a *sobiky* with a new clove seedling and one coffee seedling to plant. It has taken him only four days to clear what the forester expertly sized up as 32 ares, or 0.32 hectare—on a 50 degree slope![10]

I thought it might take me four days even to climb up it. Fallen tree trunks, interlocked fallen twigs, and underneath all greasy clay. Fortunately one great tree had been wrapped in lianas, so I did the Tarzan bit of rappelling up a liana—except with near-vertical ground below, not a standing tree.

Then, from the top, a panorama of sea, white beach in a crescent cove, the paddy fields below half-grown with emerald rice. (An overworked adjective, but what do you say about a jewel-bright green?) An isolated giant tree below the clearing drew a flock of vaza parrots—grey, awkward, social, drying elaborately fanned wings in the sun. *Vaza* means 'parrot,' but the word is very like *vazaha*, which means 'a foreigner.' When you come into a Malagasy village, all the little children rush out shouting *Vazaha! Vazaha!* Now the parrots were doing it. I grew distracted making up a jingle:[11]

> The parrot is a silly bird
> She squawketh as she flies.
> She feedeth on the toughest nuts
> That other birds despise
> She liveth many decades
> Tho' little doth she learn

---

10. *coupe-coupe*: machete; *sobiky*: straw basket.

11. Lesser vaza parrot: *Coracopsis nigra*. Cf. the ancient English folk song, 'The cuckoo is a pretty bird.'

> But every time she squawketh
> Vazaha! Vazaha!
> My ears begin to burn.

That's actually a canard about the parrots learning little: when a whole flock meets up in a tree, they hold what for all the world sounds like conversations. Another tree held bulbuls, warbling through their yellow beaks, and there was one gleam of iridescent blue from a sunbird with his brown mate.[12]

Above us on the hill behind rose the tiers of rainforest, with its cool darkness and the smell of damp leaf mold that intoxicates biologists the world over. At the clearings' edge, a ramy stump a whole meter across gives way to the desolation in front.[13]

All right, show us how it's done. Martial was understandably loth to demonstrate. Ratiarson, the forest agent, took me aside and said don't talk about authorizations, and certainly not in French. I asked if Ratiarson could ask Martial why he chose to cut primary rainforest on this steep slope, but that didn't get translated—I just got the forester's opinion of the insanity of the choice.

Martial finally agreed to pose for photos, whacking at an already severed tree that wouldn't fall down, held up by its treetop lianas. This took Frans an hour to shoot. Martial will have to wait a year for the trunks to disintegrate a bit before he burns and plants—perhaps more, because this is a sunny slope and the trees rot faster in forest shade. I'm glad he thinks it's sunny. It rained and rained for us.

Why did Martial do it? We went to his house afterwards. He's not rich enough yet to get married but he has a pretty young fiancée and a son of three. The boy is one of the worst-nourished kids in the village—not only potbellied,

---

12. Madagascar bulbul: *Hypsipetes madagascariensis*; Souimanga sunbird: *Nectarinia souimanga*.

13. Ramy: *Canarium madagascariense*.

like so many, but with the blond temples of kwashiorkor.[14] Martial has a salary—the minimum wage of 21,000 francs (= $33.33/month) as an assistant in the coffee nursery. However, he has no land, only a rice field 30 km from here that his father gave him. Relatives tend it; he gets only some of the rice. The clearing was his bid for freedom. But he could have asked permission. And perhaps even received it.

Saturday, Sept 28: Ambanizana Village

Ambanizana has 91 households, according to the careful census notebook of the village president. The school has 150 kids in three classes from the 91 households—that's about half the kids in town. For each child the parents pay 160 francs/month × 70 lots of parents, or a total of 11,200 francs/ month (= $17.77). The majority of the kids look to me to be potbellied with malnutrition and worms.

This is in a town with good rice fields, fish in the front yard, and saleable cash crops. The wide main street has maybe twenty houses a side, most with corrugated iron roofs. There are three traders with multi-roomed houses who sell candles, matches, cooking oil, kerosene, rum. There are cattle corrals fenced in bamboo, with black- or toffee-colored humped zebu bulls, a rope wound through their noses and behind the horns. One elaborate house has even got barbed wire! And lush fields of rice down to a wide river flowing from a panorama of forest-clad mountains, cloud in wisps like veils over their faces.

My chief contact is the *vulgarisateur de café*, the extension officer for coffee and clove seedlings. He speaks easy French, writes a neat hand, and showed me his geography textbook from 5th grade so we could see where I live on the East River in New York.

---

14. Kwashiorkor: protein malnutrition.

We wade the shallow half of the Ambanizana river and take a pirogue for the last 50 yards.[15] The nursery is laid out with shaded pavilions, perhaps 100 seedlings under each. The plants are in plastic bags (imported), green and healthy. They come free to the *vulgarisateur*, but he sells them at 10 francs/coffee plant and 15 francs/clove tree. Little clove trees are pompoms, like trimmed box trees but taller—new leaves are yellow frosting on the tree. Clove trees are adorable as babies as well as half grown—darker green cores, then yellow leaves, then sprigs of pinkish-rose. Clove buds are crimson. That's what we eat.

It is strange working with Frans Lanting. He can be infuriating. He spotted a picture-worthy family having lunch and made them pose for the next forty minutes, the two-year-old holding a spoonful of rice stuck half-way to his mouth. That's nothing: he spent three hours on a leaf-tailed gecko, its red mouth agape. The problem is that he is an artist of truly extraordinary vision, and nobody can get in his way. Not even a hissing leaf-tailed lizard. He rewards his (human) subjects with Polaroid portraits. Everywhere he goes, forty or fifty of the village children skip behind him giggling and screaming with delight. A pied piper in khakis with an oversized camera case.

Saturday, Sept. 28: Party!

Party! The moon rose full behind the highest peak of the Masoala Peninsula. A few white clouds on the top like an orator's head of white hair, brilliant in the moonlight. Or like lines of breaking waves, gleaming in the night, breaking silently and still on the mountain crests.

---

15. pirogue: the word for a dug-out canoe throughout francophone Africa.

Main street roofs gleam—full moon is the only light that flatters tin roofs. Coconut and banana palms shimmer as though they were carved in silver foil.

Main street is full of gleaming people. White *lambas* (printed cotton cloths) and blankets wrapped round, and little children in a shirt with one button and the shorts below the belly. All whitened with moonlight. The drummer arrives: the agricultural assistant in white shirt and trousers, his drum a tree slice covered both ends with cowhide.

The president's aide calls the throng with the official trumpet—a conch shell, blown to the rhythm: COME ON ALONG! COME ON ALONG! ALEXANDER'S RAGTIME BAND! (Its lowing tone called the people earlier to announce the dance tonight, and the next day serves as church bell.)

The children begin first, led by an adult—'Didier! Didier! Didier Ratsiraka! Didier is our father and our family!'[16]

The drummer is encouraging them with a beat going *Petraka—Petraka—Petraka*. The beat like a runner's heart which says faster—faster—faster. Only thirty or forty of the children know the songs from school, with movements to match, the six- and seven-year-old girls jigging and raising their arms and twirling in unison. Approximate unison—a four-year-old little sister who's also in the front row provides counterpoint.

The drummer whips them into marvellous rhythms (he's also the Catholic youth group leader). A lead singer takes over, piercing soprano, about harvesting hectares of rice for the revolution, as the children back her chanting *Tsigomaha, Tsigomaha*, hypnotic as Wimoweh.

But all this was warm-up. A great shout as the *betsa-betsa* came in on the shoulders of three strong men. Two whole

---

16. Didier Ratsiraka, president of Madagascar 1975–93, 1997–2002.

plastic jerry cans and a demijohn in burlap, with a great tin washbasin to dip out of. *Betsa-betsa* is made of fermented sugar-cane juice, strong as wine, not distilled. Not sour, like African millet beer, but dry and very sweet together. The name means 'you swallow large bowlfuls' but the gathering was held in check by having only two small bowls between them, one of them mine. The servers dished up a bowl at a time and passed it into the throng, not excluding tastes for the children—it took at least half an hour for a round. One elderly lady brought her stool down and put herself in front of the inner ring. When a wine server bent down obscuring her view, she smacked him smartly on the bottom of his patched shorts.

The drummer, refreshed, set to, and a few older girls began to dance. The *commerçant*, the wholesale supplier for the shops, stepped in. In this solo style the women scarcely move but their heels hit the ground to the drumbeat so hard the ground trembles. The man danced far more suggestively, arms out at girls' waist level, knees bent, bottom out and side to side.

The drummer asked for rum, knocked back half a glass like Russian vodka, picked up his drum and danced to the beat, kicking wildly. He then intercepted a glass meant for three woman dancers. The last was his downfall. Literally, taking Frans's tape recorder with him. He staggered out of the crowd and started a fight on the outskirts until half the men chased him down the street. Ratiarson the forester was disgusted at this behavior in a government official.

A second less fevered drummer took his place. The dancing went on. And on. At last they came to ancient songs—'when my mate is gone, how lonesome, lonesome.' The lead singer took over that one, simultaneously singing, dancing, nursing her baby.

When the moon stood at 11 p.m. over Main Street and the rum was gone and the *betsa-betsa* too, old forester Ratiarson and I went home. Frans said the dancing went on till midnight, everyone in it by the end, in lines and in pulsating circles, adults on the inside, children in a wider circle around.

But I found the beach deserted at last with no one likely to come, the waves breaking luminous on warm sand.

So I went swimming.

THREE

Burning baobabs; death of children

Madagascar's west coast is very different from the rainforests of the east. A 6–8 month winter without rain means the forest is deciduous—brown and partly leafless in winter, buzzing with birds and mammals and insects in the summer. Lemurs have diverse ways of surviving the winter, ranging from actual hibernation (fat-tailed dwarf lemurs) to subsisting on low-energy mature leaves (lepilemurs) or tree-gum (fork-marked lemurs), even to actual carnivory of smaller lemurs and birds (Coquerel's mirza lemurs).[1]

Not surprisingly, the western forests are very slow to grow back once cut. Have you ever seen a poster of Madagascar showing the magnificent baobab alley of Morondava? A double line of towering trees with a road between, variously crowded with oxcarts, tourist buses and film crews? The flat land on either side, though, was once forest—all but the baobabs themselves have gone.

From the early 1980s, a Swissaid scheme attempted to find a way to log this kind of forest sustainably, based on good research. This was in the Kirindy, south of the Kirindy river. Ten years of intensive effort eventually showed that there is no commercially sustainable use, though the Swiss are still

1. Fat-tailed dwarf lemur: *Cheirogaleus medius*; lepilemur or sportive lemur: *Lepilemur*; fork-marked lemur, *Phaner furcifer*; Coquerel's lemur: *Mirza coquereli*.

trying to engage the people in participatory conservation. Trees are just too slow-growing for forestry in this climate. Some only seed at six- or ten-year intervals after a cyclone.[2]

As for the local *tavy*, to paraphrase what the Swiss foresters told me, 'After burning the forest the ground is sown to maize without further preparation. A crop is grown for 1–3 years on the same area. There may be a return to the cleared field after 5 years. The second burning eradicates all seedlings of forest trees which may have sprouted and finishes off any large trees that survived the first clearing. After the second treatment, the soil is barren. The worst area is around the village of Beroboka where people have savaged the forest with an apparent will to destroy it.' Beroboka lies in the middle of a vast swathe of the forest cleared for sisal in the 1960s around an ever-flowing river: a failed plantation.[3]

Just after the Duke of Edinburgh's visit to Beroboka and his speech to the conference on Conservation and Sustainable Development, photographer Frans Lanting and I also turned up at Beroboka to write our conservation article for the *National Geographic*. We stumbled into a slice of village life I never hope to see again—what happens when poverty and polluted water combine to murder children. Enough to forever douse the naive outsider's view that village life is romantic.

---

2. J.-P. Sorg, J.U. Ganzhorn et al., Forestry and research in the Kirindy Forest Center/ Centre de Formation Professionnelle Forestière (2003); C. Dirac, L. Andriambelo et al., 'Scientific bases for a participatory forest landscape management in Central Menabé' (2006); J.U. Ganzhorn, 'Cyclones over Madagascar: fate or fortune?' (1995).

3. The plantation was old Monsieur de Heaulme's pledge of faith in newly independent Madagascar. After he bought it, just before 1960, when other companies fled, he felled a huge hole in the forest and planted sisal. The plantation functioned until 1979, when his son Jean de Heaulme was briefly jailed by a government determined on further nationalization. A. Jolly, *Lords and Lemurs: Mad Scientists, Kings with Spears, and the Survival of Diversity in Madagascar* (2004).

October 13, 1985. Beroboka. The plantation manager fears poison.

Landed just south of plantation house, with gleeful children running to watch us arrive in the Amoco helicopter.

We sit in the high-ceilinged concrete dining room of the old plantation house at Beroboka. Table holds a kerosene lamp which lights the faces, while a mosquito coil on the floor partly protects the ankles. The faces lean into the light. Ferdinand de Campe is the plantation manager for the de Heaulme family, with black brow and mustache and one red vein standing out in his eye. He is voluble, intense—and vulnerable. He talks and talks and gestures, as all do who have guests at last on a lonely plantation in the bush. He is indeed committed to the preservation of the forest, to the point of having arrested the chief local rabble-rouser of people to destroy the forest. He had to drive the police to arrest the man in his own car. Even then it took eleven of them with guns to make the arrest, confronting one fellow with a stick.

'Rabble-rouser' isn't quite the right word. This was the local radical politician, who argued that the French de Heaulme family had abandoned the land and the people should take possession—especially of the forest. They got him arrested, anyhow.

Then de Campe held *kabary*, a traditional consultation, with all the locals, including the President du Fokontany Beroboka Sud (Beroboka community), Étienne Gilbert. Everyone agreed that de Campe should ask the de Heaulmes to uproot the bulbous bases of old sisal and plow and harrow permanent land for maize and peanuts. De Campe has the seed to sow, but no tractors capable of preparing the land. The locals feel betrayed, and so indeed does de Campe, because the season when they must work the land is right now. The fokontany, seeing no tractors, has applied for the right to clear 1,000 hectares of Water and Forest Ministry forest land instead of the sisal-studded ex-plantation. The permission was granted a day or so ago, Oct. 10 or 11, then apparently rescinded October 13. Now pending developments.

In this highly seasonal climate they must clear now to burn later, and plant late November or early December. Otherwise they miss the rainy season.

Meanwhile de Campe has his food and drinking water watched, and goes armed in the forest. There is not only the traditional Beroboka village of local Sakalava, but a whole group of Tandroy, brought here from the south as sisal workers when the plantation was running. They have settled in a kind of forest-edge linear slum. *'Ceux-la, je ne les ai jamais apprivoisés. Ils m'en veulent à mort, car j'interdis les brulis.'* (Those people, I have never tamed them. They want me dead because I forbid burning forest.)

Our own party includes ornithologist and zoo director Georges Randrianasolo. Georges was the man who accompanied many many expeditions, including Dave Attenborough's 1961 BBC film series *Zoo Quest to Madagascar.* He has grown thinner than ever as he ages, but no less voluble, like the parrots themselves. Photographer Frans Lanting fingers and twists the corner of his mustache or the near corner of his beard. Why is his hair so straight when his beard is frizzy? And why do I find a hair color of mid-yellow brown is ill-matched to an icy blue eye? Anyway, since he can't follow most of the French as Georges and de Campe get deeper and deeper and more repetitive about what's wrong with Madagascar, he isn't looking grim as usual. I'm feeling grim instead, at the endless conversations which start *'Avant...'* *Avant* means before... before Independence, before the French left. Everywhere people compare present poverty with *'Avant...'* but of course they never say the rest out loud.[4]

4. Tsimbazaza Zoo, in the capital, Antananarivo. The zoo is in a lovely botanical garden, founded by French scientists in 1927 beside a traditional sacred lake. It has its

Finally, Hubert Randrianasolo, Frans's assistant, tall with wavy hair, and a slightly plump smooth face. Hubert makes you think of flesh and skin and hair while Georges is bone structure and bright eyes. Hubert's father was a noted artist. His grandfather was an MDRM politician who died penniless in prison, and wrote to the family never to become politicians again. Therefore intelligent, sensitive, tolerant Hubert owns cars, and is determined to succeed as a businessman. Probably only an artist's son could tolerate work with Frans, and recognize that Frans's arrogance is not for his ego, but for his art.[5]

Frans and Hubert go out to photograph the hand-raised fork-marked lemur till midnight. That means photo-ing an animal that keeps crawling down the back of their shirts instead of posing on the tree, all amid mosquitoes and lightning and lowering heat, and in the background village voices wailing the dead.

October 14, morning: A giant jumping rat problem.

Frans and Hubert up before dawn to drive to a spot where Frans foresaw a photograph of a huge red sun rising between two distant baobab trees. Frans got his photo.[6]

I wish I could like this forest better. It rained last night. The mosquitoes fasten like Sartre's Furies. Sartre really missed out making the Furies that punish sinners into 'The Flies'—they're actually mosquitoes.

The baobabs are beautiful, but the stuff between mostly looks like second growth ready for the *coupe-coupe*.

---

ups and downs, depending on foreign aid and a government policy that generally treats keepers as no better paid than road sweepers.

5. Mouvement Démocratique de la Rénovation Malgache. A pro-independence political party of intellectuals founded in 1922, and forcibly dissolved in 1947 during the French repression of Madagascar's 'insurrection' or war of independence.

6. That photograph is one of the iconic Madagascar images. Many other photographers have copied the idea since then, but no one has Frans's precision of dramatic vision.

And of course this is the wrong season. I keep reminding myself October here is the equivalent of the February thaw in my home town of Ithaca—bare branches where the sap is beginning to run, but it's impossible to imagine summer-time. De Campe is lyrical about its beauty in March when the summer heat has abated but all the woods are still green and plumed.

Even now the birds are beginning to sing. I see *Coua gigas*, a giant coua, and *Coua cristata*, the crested coua, and a third, new to me—intermediate size, grey and pointy-tailed like *cristata*, rufus breast and ground-running like *gigas*. Georges says *C. reynaudi* or another.[7]

Frans and I go down to the Kirindy river (wholly dry, of course) where Georges has been setting traps for the giant jumping rat, *Hypogeomys antimena*, in hopes of bringing some back to Tsimbazaza Zoo. Giant jumping rats are among the world's cuddliest-looking rodents: long pink ears like a young rabbit, pink woffly nose, upright sitting posture with the appeal of a pet gerbil. And they do hop, even at 10 in the morning when the road shimmers in the heat.

At noon, as we lunch off rice and yet another tin of corned beef, young Silvio Flueckiger, Urs Rohmer and Dr Sorg of the Cooperation Suisse drop in. They are doing lovely scientific work, planting seedlings in different months and then replanting them in the woods.

Their big problem seems to be the giant jumping rats. The seedlings, which have been well watered in the nursery, are of course greener than any other dry season stuff. With meter-high woody sprouts, the something comes back three nights running. The first night it debarks the sapling, and the second night cuts it down and eats the leaves, and the third

7. The couas belong to an endemic Malagasy bird subfamily, the *Couinae*, related to cuckoos.

night it digs up and eats the roots. It adapts beautifully to their 10 meter spacing out.

I congratulated Silvio warmly on his providing food for one of the world's rarest mammals. He looked exceedingly mournful.

October 14, afternoon: Funeral of Monika, aged 3.
The old poacher, the 'Père de Blaise,' whom we'd thought might show us a lemur trap, has had two grandchildren die in the past three days—one here and one in Belo sur Mer. The wailing and chanting last night were to drive out evil spirits from the village. Three children have died here in a few days, and more are sick.

A funeral procession passed the plantation house. Hubert ran to ask if we might photograph. For a photographer, a village funeral was an unmissable opportunity, as long as the family agreed. Amid confusion at the sudden interruption the father said yes. Frans walked with the men. The rest of us—Hubert, Georges, me and about four local villagers followed in a 2-seater jeep. Hubert drove back for rum for our contribution—it is *fomba gasy*, Malagasy custom, for mourners to contribute rum. We reassembled with the funeral party in the woods.

The married women knelt and squatted in a little circle, just inside the woods, around the coffin. Younger people of both sexes stayed across the road, well away. It was only a little coffin with a peaked roof like a house. They had a cross: two planks nailed together, with pointed top. Burned in angular letters:

MONIKA
MATY
10–10–85

Monika had been three, or three and a half. The box was wrapped in bright *lambas*, the outer one Independence

Anniversary 1985, red and yellow and green. It was all a bright, bright scene—the gay little red and green coffin, the women in multicolored *lambas*, black Sakalava skin and white-barked western trees. They sat clapping and singing repetitive minor chants till one cried *Sifaka!* Behind, from a tall bare white emergent, a troop of white sifaka soared against the sky with their ballet grace—10 feet down to a lower tree and onward, branch after branch till they disappeared.[8]

The women began to pass home-made spirit—*toaka gasy*—around in a green plastic mug, bright as the *lambas*. A few stood and danced, the usual hip-swinging stamping in time to the clapping. One woman shouted angrily at *vazaha* attending (Frans snapping pictures with apparent phlegm). I looked at Hubert, but he kept his cool, declined to translate and figured internally when and how he might extricate us, if necessary.

A man came from the woods, sat in the circle of women, and said in ceremonial sing-song that the grave was nearly dug. He was passed some *toaka*, and returned to the woods. At last all the men came out (there were about 20 married men and 20 married women at the funeral). The whole group turned in procession down the path.

Only then we saw what the graveyard was. The path was lined both sides with graves. Many fresh—one child's 12–9–85, just a month ago. Each child's grave had a little hat on the cross: cloth store-bought hats, not local straw. Ten or fifteen were fresh, new, this year's hats. Little graves with new hats. Many of the graves were adult though. Monika's was inserted between others, perhaps 4–5 foot deep in the sandy soil.

---

8. *Lambas*: wrapped cloths. White sifaka: *Propithecus verreauxi*.

They made a little fire in the woods to burn her few possessions. They lifted the coffin lid, and folded the three bright *lambas*—white, blue and white, red and green, and tucked them over the body. All the while the *toaka* cups passed round, as people grew loud and argumentative and laughing. I quoted my father's remark about wakes growing jovial as wakes do, but Hubert was shocked—Not on the plateau, he said. First the father of the family, then later the women, offered me *toaka*. Even a sip hits you sharply under the left eye, with a twinge of pain up into the sinuses. Perhaps more of it is anesthetic.

And all the while the sun filtered down through the branches in the glaring colors of afternoon. Even the pile of yellow sand looked like something seen in Kodachrome in the *National Geographic*.

At last the women each stepped to the sand pile. They turned their backs to the grave. With a quick backward scoop of the left hand, they tossed sand onto the coffin, already running away, so that death should not follow them. Then the gravediggers set to with shovels.

The procession started back to the road. A few final ceremonies remained: washing the hands in water the women had carried on their heads in red and blue plastic buckets. They had started to dispute by now, angry with each other and with us. Hubert reached his cut-out point, and told us to leave.

October 14 evening. Death of lemurs.

Last night Frans and Hubert and Georges and I had an intense ethical discussion around the pressure lamp and the ankle-height mosquito coil. Should Frans photograph lemur poaching? Would we be sanctioning what was both illegal and, to us conservationists, deeply immoral? I am afraid I was the most hard-boiled. It happens anyway. It's part of local culture. We would just be recording it. An old local poacher agreed to help us, but did we in fact share responsibility for the deaths of the particular lemurs he proposed to catch?

So this morning our ancient trapper and Georges and Hubert and Frans and I climbed in the jeep and drove to the Kirindy river. We started walking down the sandy streambed under the same parched woods. My ethics, or just plain squeamishness, evaporated. We weren't egging on the old man to build a new lemur trap. He just took us to the one he'd already built over the past two weeks.

The trap was a runway of two cut trees, in all about 20 ft long and 5 ft above ground. The saplings rested in the crotches of live rooted trees. He'd cut all the other bushes for several feet around and lashed the runway together with strips of *hafotra*, or fibrous bark. Spaced along the runway were eight actual traps. Each had a trigger: the bent top of a still-rooted sapling. The sapling looped into a strap of more bark. He first bent the saplings into the straps. Then he cut 4 ft lengths of stick, and put them horizontally to hold the straps apart, top and bottom. Then he tied what looked like string to the triggers, in a running noose.

Georges promptly showed me how to make the string. You slice off more *hafotra*, split off half-inch-wide strips, split that in two, sit down on the ground, and pull your blue jeans up from your safari boots to reveal a length of bony skin. '*Je reviens à mes ancetres*' (I'm turning back into my ancestors) grunted Georges. In a second's rubbing on the shin the *hafotra* is transformed into a length of twisted string, as regular and strong as commercial nylon.

'Lets go see what the old codger is up to,' said Georges. The old man was posing with his spatulate fingers at the traps while Frans and Hubert deployed silvered mirrors in the forest to light up his profile.

Hubert seemed to know surprisingly well what was coming next. At last he admitted he'd made just such traps as a boy in Morondava. In fact, everybody can make traps like that. He didn't actually say he'd caught any lemurs, but certainly plenty of birds!

I could see, as if it was happening, how a female rufus lemur would be garroted, dangling with broken neck from the sprung trap, her fur red in the spotted sunlight and the surprised look of her white eyebrows now set in the surprise of death. Or worse, a male caught round the waist, screaming small lemur screams until the old man walked back to finish him off—or the crows and kites got there first.

Even counting an old hunter's time as worth nothing, the effort can hardly be worth the meager return in protein. People just like to hunt. The incalculable part is finding your lemur troop and learning its routes. This season is the *soudure*, the hungry season between harvests for lemurs as for humans. There is almost no fruit left, and no new growth. The rufus lemurs react by doing nothing all day, and sortie-ing for an hour or so at dawn and dusk. This means your watching must be only at dawn and dusk. And the region is 10 km from the old poacher's home!

Then building the trap takes a full day.

Then finding bait. You go locate a lepilemur.[9] If its nest-hole is accessible, you catch it with a brush-ended stick twisted in its fur. If not, cut the tree down. We stepped over a tree that fell across the Kirindy, about a foot in diameter, felled to get one lepilemur. You kill the lepilemur and take out its intestines, which you smear on the runway: this is the favored bait. Flashback for me to Jersey Zoo, and that Mayotte lemur female with the mouse intestines

9. Lepilemur are sometimes called sportive lemurs. An odd joke, as they sit immobile for most of the day and night, conserving the meager energy from their mature leaf diet.

dangling from her jaws, while her yearlings jumped and snatched at them. 'I never heard they liked lepilemur tripes!' exclaimed Georges. 'It's new to me!' So much for vegetarian scientists.

Fortunately we were too late to go through this and contributed bananas for bait instead.

October 14, night. Death of children.

So home. Dinner with de Campe—tomato salad, and chicken in fresh ginger sauce, and sliced green mangoes, served by smiling Mama in that dismal dining room. And me all the while hoping we'd not have to do the next bit with the lemur trap.

We didn't. Hubert and Georges went off to the village to see if there would be more spirit chasing for the next child who died just after Monika's funeral. Came back an hour or so later shaking.

The child's family were Christians. The older brother, even with tears in his eyes, said they'd hold a traditional rather than a Christian wake, since we were here to see it. Georges and Hubert were sat condoling with the family. Meanwhile a *charette* (oxcart) is bringing down the body of another child who died in Belo. (Even the body of Monika was perceptibly rank. I don't like to think about a body that spent two or three days in the sun in an oxcart.)

As Frans and Hubert sat with the family the news came that yet another child was dead. One old man has lost three grandchildren this week. The family excused themselves—could we please stay away?

Hubert nearly cried himself. His own son is 6, his daughter 4. He said he tried to be tough, but he couldn't be with children. He couldn't stand it. Was there anything we could do to help the family—or the village?

October 15. Fancy ice cream cake and dirty water.
Dawn. Packed. Down that blasted road to the Kirindy. Into the woods to the lemur trap. No lemurs, thank God.

Another child has died. That makes three last night.

And then through a guarded gate and chain-linked fence into Alpha 2, the Amoco Oil base camp.

I thought it would be a town. Instead it is linked trailers made into offices, and a giant prefab hangar. Everything is temporary, ready to move to New Guinea or Chad if there's no oil here. And everyone speaks American. I put the case for airlifting a doctor to Beroboka village this afternoon.

And then they airlifted me to Tana.

And at 8 p.m. I went to dinner with the Greers (head of Amoco in Madagascar) with crystal glasses on the table and a Pakistani pierced wooden screen behind and a silk Persian rug on the wall. Chatted with the Egyptian ambassador, who had to tell me about the current international crises from scratch. Funny that an Italian cruise ship can be hijacked by Palestinians and an Egyptian plane hijacked by Americans in the past week without people in Beroboka noticing.

Also the head of Fraise & Cie, who commands the vanilla market, and his wife who wears silver armored eye shadow. Also an Englishman from Mombasa who does all the heavy marine transport this side of the Indian Ocean, and laughed at my jokes.

I kept myself together till the ice-cream cake arrived for desert. The cake was left from the Greers' 2-year-old daughter's birthday party the day before, about the time we were burying Monika.

So I gave Eva Greer the whole story. And God help me I was pleased when she started crying into her ice cream. I guess Beroboka will get its medicines—and maybe some other villages too.

Home to bed.

October 16.

I guess this was Sunday, sleeping and washing underwear and trying to write all the feelings away. I can start to Fort Dauphin tomorrow with a cleaned-out brain, and my hair washed.

Dinner in the Colbert: *salade verte, magret de canard à l'orange, pommes nouveaux, fromages de France*, and two bottles of beer.

October 17.

Update on Beroboka. The Amoco doctor flew in, in a big 212 helicopter 10 minutes before Frans, Hubert and Georges had planned to leave. They trooped off to the local nurse (whom they'd already met) in the Poste Sanitaire. This is a hideous mud hut, filthy. The new dispensary has been built but never opened.

The nurse and the traditional healer and a midwife who's a close colleague and the Amoco doctor all agreed on diagnosis. It's not measles, but diarrhea brought on by drinking the polluted water left at the end of the dry season. The cure is simply oral rehydration: a liter of boiled water, six teaspoons of sugar and a half-teaspoon of salt to make a solution no saltier than tears, continuously spooned into the sick child's mouth. The sugar makes the gut absorb enough water to counter diarrhea. Unicef is now promoting oral rehydration all over the world.

One mother had even taken her baby to Morondava, had the same diagnosis, but was 'too poor' to buy salt and sugar, so the baby died. (I suspect she didn't trust the simple remedy, hoping for a curative injection.)

One family whose child died on a Sunday made funeral preparations on Monday, and waited through Tuesday, Wednesday and Thursday, which are *fady* (taboo) for burials. The whole area stank, of course. Meanwhile they killed two zebus to feed the assembled family. As Hubert said, 'When you're that poor, you really need a party at times.'

Anyway, Frans and Hubert rushed round and bought 8 kg of sugar and 2 kg of salt. At Hubert's suggestion the nurse mixed them in the proper proportions so they wouldn't go into coffee or rice. They held a *Kabary*, assembling all the mothers they could find. Frans waving his tablespoon and his teaspoon, and everyone explaining and explaining.

FOUR

David Attenborough, Madame Berthe's mouse lemur, and school among the baobabs

Much has happened in the Kirindy and Beroboka region since that week in 1985. The Swiss Forestry scheme finally closed. Their ambitious hopes ran into too many roadblocks. To the north-west is a Unesco World Heritage Site, the spectacular stone pinnacles of the Tsingy. The baobab forest of the flatlands are partly protected instead as the *Aire Protégée du Menabe Antimena*, established in 2006.

The Durrell Wildlife Conservation Trust (formerly Gerald Durrell's Jersey Zoo) has tried to turn villagers from occasional destroyers to full-time protectors of the forest. Every year Durrell runs competitions between the villages to see who has done best at conserving their nearby forest patch. The people walk together through each area. They count new *tavy*, freshly cut stems, lemur traps and also birds, mammals and giant jumping rat burrows. Winning villages get substantial cash prizes amid a festival party. However, this is not enough to provide livelihoods for all. In Beroboka the abandoned sisal stumps are finally gone and all those fields turned into crops: lush irrigated rice—but again, there is not enough for all, with the growing population, so *tavy* continues.[1]

---

1. J. Ratsimbazafy, L.J. Rakotoniaina et al., 'Cultural anthropologists and

In the heart of the Kirindy forest is a new research station run by the German Primate Center of Göttingen. Eager young German researchers live there in tents and monitor the mammals. In 2010 a BBC team arrived to film their natural history series on Madagascar.

This, of course, will be the Western impression of Madagascar: gloriously presented in the magisterial voice of David Attenborough.

Oct. 7, 2010, late evening, Antananarivo. The Crew.

David Attenborough rises to hug me and say hello and cheerily offer beer. He settles into beaming welcome and dry humor, maintained through all of the quiet moments of the next few days.

The crew: Sally the producer (blond, curly, just about 40? Or younger?); Ian, cameraman, intense, dark-haired, cleft chin; Grahame, retired sound man willing to come back to work again with friend David, curly gray hair about as long as Sally's, red face and perpetual smile. Also Emma Napper, who got me into this. More on Emma later.

We drank our beer and nibbled cheese toasts till 9.30ish, when Sally pointed out that we had to leave for the airport at 4 a.m. David didn't seem tired.

Oct 8, 2010. No memory of plane trip to Morondava.

I didn't wake up until we disembarked at about 8 a.m with our twenty-odd pieces of luggage, mostly the silver-shelled stuff that holds camera equipment.

The road to Kirindy—the first 20 km was once tarmacked, now deep potholes, so the usual sashaying from

conservationists: can we learn from each other to conserve the diversity of Malagasy species and culture?' (2008).

side to side finding bits of dirt to drive on. The next 45 km dirt plus potholes—none of it fast enough to develop corrugations. The famous baobab alley, and a whole lot more baobabs, and a recently burned area, and then endless scrubby Kirindy forest with baobabs sticking up, the other big trees having gone.

On one of the mini-detours David and I simultaneously clapped our hands over our ears. We each removed one hearing aid and then the other. We sheepishly looked at each other when we realized that the beeping was the Land Cruiser backing up.

Kirindy is now a tourist camp as well as a research camp. They have used the BBC filming fee from last year to build two new huts with matting walls and indoor flush toilets! And showers. David got one, I got the other. The rest are in older shared huts with an outside long-drop.

Over lunch, it turned out that Emma spent three months here last year. David at first amazed, not so much that the BBC funded her but that she thoroughly enjoyed the stay. It turned out her crew wanted the change of seasons. So much happens and is over within a day or two, once the rains set in, that they couldn't risk leaving. Notably little frogs that turn bright yellow and mate for a single day in the puddles of the first rain, and then disappear, camouflaged, for the rest of the year. Ian also amazed but more scornful than humbled. He is not a wildlife cameramen but a generalist mainly there to film David. He expected plumbing.

Emma rushed round—assigning rooms, sorting out where to do pieces to camera, attempting and failing to get phone connection to stave off a party of other tourists, warmly greeting the scientists. Melanie Dammhahn is here to study the Madame Berthe's mouse lemur; Mia-Lana Lührs to finish her Ph.D. on fosa. They are partners: Melanie the dark smiling, dimpled, in her trademark black tee shirt, Mia-Lana the beautiful blonde with freckles and a natural curl in her shiny cropped hair. Half a dozen more researchers are here from the German Primate Center in Göttingen—they have even more Spartan accommodation than us, and, it is rumored, an even worse long-drop.

Emma seems much more like a scientist than a media creature—greeting people, sorting things out, knowledgeable... A small endlessly eager and cheerful woman with a long brown pony tail, looks about 25 though she's all of 32. Turns out she has a Ph.D. in entomology. She got into the BBC because she'd done her doctorate on the large blue chalk butterfly, which tricks ants into rearing its larvae, and advised on a film sequence. She knows exactly what needs doing in the Kirindy because she set it all up. But she says of course she understands why she is not the producer—they expect someone with more general experience to be directing David.

This leaves Sally with the worried expression that I associate with film directors and have done ever since my first BBC film in 1980... a furrowed brow and permanently abstracted air. I said as much to David, who said, 'Well, the buck stops there. On these shoots, I never argue with the director. I am simply an extra piece of luggage. They wheel me on when wanted.' Of course David has probably directed more films than Sally has years of life.[2]

Cameraman Ian wishes to decide on every shot himself. Nothing will go right unless he is properly consulted. Emma knows everything about the site and the shoot; I think I know everything about Madagascar; David really does know everything about everything but refrains from saying so. I wouldn't be Sally for anything.

2. A. Jolly, 'Tropical Time Machine' (1983).

Thank goodness for ebullient Graham! And the fosa[3] that appears at lunchtime and curls up on a table in the next building. Graham whips out his still camera—he ends with a lovely shot of the fosa walking, the only one I have ever seen which makes it look like what it is: a huge carnivorous mongoose, not a cat at all. But Graham got between the fosa and the way it wanted to go. It advanced on him growling. He turned tail and came back claiming to have been nearly attacked by a cross between a kangaroo and a leopard. He is ignoring the fact that it only stands about knee-high—he's not messing with it.

Mia-Lana Lührs has worrying population statistics on the fosa. In three years of trapping she has caught and identified only 23 males and 9 females. The females are territorial. Probably no more than she has already seen in the whole forest! And she has seen and filmed and written about a cooperative hunt when three together treed a sifaka and cut off its escape routes one by one until it was too tired. Then they caught it. It turns out that males that associate with each other grow twice as big as the females, up to 11 kg instead of 5. Solitary males are female-sized. A weird social system. Maybe Graham is right to go round wielding his sound boom.[4]

In fact there are two camp fosa, brothers called Itchy and Scratchy who hunt together. They are the camp's main tourist attraction. The tourists by now have taught them that (1) people may offer food, and (2) people are subordinate creatures who run away if you growl at them. The scientists

3. Fosa: *Cryptoprocta ferox*, a puma-sized Malagasy carnivore, related to mongooses though cat-like in appearance. Madagascar's largest predator. *Fosa* is the Malagasy spelling, pronounced 'Foosh.' It is often spelt *fossa* in English, but this confuses it with the much smaller carnivore *Fossa fossana*.

4. M.-L. Lührs, 'Simultaneous GPS tracking reveals male associations in a solitary carnivore' (2013). M.-L. Lührs, M. Dammhahn, et al., "Strength in numbers: males in a carnivore grow bigger when they associate and hunt cooperatively' (2013). M.-L. Lührs, 'Polyandrous mating in treetops: how male competition and female choice interact to determine an unusual carnivore mating system' (2014).

are terrified that they may actually learn that (3) tourists are made of food, in which case the fosa will be shot, not the tourist, which doesn't seem fair.

Oct. 8. The focus of this shoot is the smallest primate in the world, Madame Berthe's pygmy mouse lemur.

Like the giant jumping rat and the narrow-striped mongoose and the flat-tailed tortoise, Mme Berthe's mouse lemurs only live in the forest of Kirindy and around Beroboka: a tiny slice of western Madagascar. An adult weighs 30 grams and could sit on your thumb. Red coat, oversized head, huge eyes, fuzzy fuzzy tail. I saw one first with Russ Mittermeier. Unlike the bigger grey mouse lemurs, which flee from cameras, Berthes freeze. They are too small to do anything else. Russ sent his pictures far and wide but neither of us had the wit then to see that this must be a very different species. Melanie Dammhahn has caught one of her study subjects for BBC stardom. She gave me a close-up glimpse of its face, calling it the cutest animal in the world. She could be right.[5]

They had to go down a side trail for David to do his piece with Melanie and let the little lemur go at its own home—no room for me. Then Anni, a Finnish student, and German Frederik turned up with a *Mirza* that had been trapped for marking! I've never seen a *Mirza* except in the depths of a dark cage in Duke. It was just a rather weasel-like lemur at first sight. Then I saw its very very long red tail, much longer than head and body. At last it had the courage to come out of its mesh cage and scuttle up a vertical trunk. It wasn't really sifak-ish, because sifakas' big hind legs propel them off in an arc up and over to land meters away. This thing seemed to just stay vertical on one trunk, then turn

5. Madame Berthe's mouse lemur: *Microcebus berthae.* Grey mouse lemur: *Microcebus murinus.*

still vertical and cling stuck to another as though with scotch tape. Eventually it moved up towards the canopy and out of sight.[6]

Apparently Berthe's mouse lemurs sleep in *Mirzas'* abandoned nests. One to three mouse lemurs may share parts of a nest with a *Mirza* in residence—even though it sometimes eats the Berthe's!

When everyone else goes to bed, Emma heads out into the woods to catch some flower insects for tomorrow.

Oct. 9, David's piece to camera: flower insects on liana.[7]

Delay. Found a fine photogenic nearby liana, but it is the wrong species of liana. Further prospecting.

More delay. I officiously say they should film the flower-mimic insects at Berenty. Apparently here they are nocturnal, hiding under bark during the day and only descending to feed at night. I say that ours hang out in a highly photogenic colony on the liana all day.

They ignore me. Emma has set things up. David uncomplainingly limps further into the woods to the chosen liana and sits in his folding chair, while Ian chooses angles and Graham aims his huge furry microphone. Insect doesn't cooperate, but hops away.

David suggests he do the piece without insect, pretending it's on the far side of the liana. All agree. He puts himself in place and launches into an almost perfect piece. It's not until the third take, though, that he pares down and sharpens the wording to his own satisfaction:

> This particular liana belongs to a species that only grows in this part of the forest. And on it lives this little insect. It's a bug. It feeds by sticking its mouthparts into the liana

6. *Mirza coquereli*, also called Coquerel's mouse lemur.
7. Flower insects: *Flatidea coccinea*.

> and sucking out the sap. It digests what it wants and then excretes the rest as honeydew, a sort of sugary liquid. And it's that honeydew, that sugar, that the little lemur needs in its diet. So Madame Berthe's lemur is only found in this particular part of the forest because of this insect and this liana. It just shows how complicated ecological connections can be, and how much you have to know about an animal if you are really going to conserve it.

At last he could sit down. And then the funny bit. I was dead wrong about filming at Berenty. Kirindy flower bugs are certainly a separate species. They are nocturnal, smaller, with a denser bustle of filmy white strands—even more emphasizing David's point about saving each ecosystem and all its parts. The bugs fan out their bustle of white sugar-strands when annoyed. Then they hop. And disappear from the close-up shot. Efficient Emma eventually had to run back to camp and put her captives in the fridge to hold them still enough for the final take.

A little book about the littlest lemur. The reason that the BBC asked me along was to read to the local school about Bitika the Berthe's mouse lemur. I tried to persuade the Ministry of Education to include biodiversity in the primary school curriculum starting way back in 1964. Others, Malagasy scientists better placed than me, have been trying for even longer. In 2005, with my dear colleague Hanta Rasamimanana and New York artist Deborah Ross, we decided to do an end run around the Ministry. We produced six beautiful, amusing books, the Ako Series, which children would want to read and teachers would want to teach. We started with *Ako the Aye-Aye* and went on to six other species living in six different corners of Madagascar. Unicef has sent tens of thousands of copies in the Malagasy/English versions out to schools, and the Lemur Conservation Foundation in Florida has printed an all-English version for children in the USA.[8]

Hanta Rasamimanana has been working beside me at Berenty since 1993. She is now far beyond me. Indeed her work shot down one of my favorite theories about lemur female dominance. Short, intense and prone to giggles, with a vision of how to bring out the best in every student. Every year she brings the top two or three of her class from the École Normale Supérieure to do fieldwork theses. By now she is dubbed 'Madagascar's Lemur Lady,' profiled on CNN, and French TF1, and even featured on a vast new IMAX film.[9]

Hanta is Malagasy author of the Ako Series. Sadly, her further promotion and research of the Ako books makes it clear that few schoolteachers are willing to tackle anything off-curriculum, anything not rote learning, and indeed are terrified of trying anything new: 'What if the children ask questions? I don't know anything about lemurs myself! I'll look a fool!'[10]

The book about Berthe's mouse lemur was a challenge to think up, because before Melanie Dammhahn's research almost the only thing we knew about this species was that it was small. My story is that Bitika, which means 'Tiny' in Malagasy, meets one after the other of the seven other species of lemur which live in that same forest. She feels smaller and smaller and smaller and *smaller* until finally she saves her mother's life by screaming at a white-browed owl, the Berthe's

8. A. Jolly, H. Rasamimanana and D. Ross, Ako Series (2012).

9. H.R. Rasamimanana, V.N. Andrianome et al., 'Male and female ringtailed lemurs' energetic strategy does not explain female dominance' (2006); B. Simmen, F. Bayart et al., 'Total energy expenditure and body composition in two free-living sympatric lemurs' (2010).

10. F. Dolins, A. Jolly et al., 'Conservation education in Madagascar: three case studies of an island-continent' (2010).

chief predator. Then Bitika feels like the biggest lemur in all Madagascar![11]

October 9, 2010. Back to Beroboka.

After lunch, when Emma had sorted everyone else out, she and I went to rehearse. The new school at Kirindy that the German students paid for isn't open because the local parents decline to support a schoolteacher. So we are off to Beroboka on the old de Heaulme estate.

Big, concrete school. 200 students, 4 teachers. This is the unruly and disappointed village which has had multiple projects to raise its hopes of work, but been repeatedly abandoned. The sisal stumps have gone, though: there are glowing irrigated rice fields by the flowing river. However, Beroboka villagers are still the worst perpetrators of *tavy* in the whole region. The baobab forest still burns.

The school director, a man, has whatever the disease is when people can't keep their eyes open and have to look through half-closed lids. (Pink-eye from the dust? Or worse? Very common.) The second oldest teacher is a Madame Teremialy—dignified, quiet, 40ish. She likes the story. We work with her to translate Bitika into local dialect. The last bit takes some explaining, when Bitika sees the owl and screams and saves her mother, but she writes her own version down in a notebook.

Emma asks if she would prefer our interpreter to tell it, but Mme Teremialy says very firmly that she is going to do it.

So back to camp with chickens in the back of the car, since we had eaten at least one up for lunch. Next morning Graham reports he absolutely had to go to the long-drop in the middle of the night. After putting it off as long as

11. White-browed owl: *Ninox superciliaris.*

possible, he went out with his sound boom in his hand to hold off the camp kangaroo/leopards in case they attacked him.

He needn't have worried. The fosa brothers were busy killing all five chickens in the restaurant chicken coop, and actually eating two of them.

Oct. 10, 2010, early morning.
Reading 'Bitika the Mouse Lemur.'

The schoolchildren were there, all in their blue wrap uniforms, on a Saturday morning at 9 a.m. They lined up before the flag and sang the national anthem. David pointed out that this is one of the few countries (the Solomon Islands being another) where the missionaries have a legacy of pure diatonic harmony. In the Solomon Islands, the traditional, pre-mission songs had the same harmony. In Madagascar, as well, traditional instruments were tuned to a diatonic scale like ours, but drawing on African drum rhythms very foreign to our ears as well as Arabic minor notes. The missionaries did complain because in their congregations few could be persuaded to carry the tune—they all kept sailing off into harmonies!

Of the kids, about 40 were small, boys and girls equal, then about 40 bigger boys and 80 big girls (up to just puberty)—easy to see as there were two lines of big girls for one of boys. Boys are too useful guarding the cattle and goats to leave in school. Only the little ones were coming to be filmed in a spare building we'd chosen yesterday, now swept clean and with three rows of benches. Emma arranged David on a chair at the back looking benevolent, Mme T. and me in front.

Picture a concrete classroom, wide open windows both sides looking out on a dusty empty tract of schoolyard (dun colored but bright in the sun), dusty blackboard with Janet

Robinson's lovely posters of Malagasy fauna and flora stuck on. Then me in my pink flowered shirt (and doubtless very shiny pink face), Mme T. in her very best burnt-sienna-colored dress and coat and shiny dark-brown face. The children with hair all beautifully braided (girls on the front two benches, boys at the back), in their royal blue and white uniforms. There was one little girl in the front row who was just as excited as Mme T. and me, eyes shining as she mimed the chorus.[12]

Mme T. was a total star! She had read and mastered the book overnight, so she told them everything in my text. We got the kids repeating the chorus Bitika felt small: *Nahatsapa tena ho kely i Bitika!* With every page, Mme T. got better and better, pantomiming all the creatures, and the kids got better and better, and we were on a roll.

Lots of unprompted questions: how many babies a year do sifaka have? How old do they live? And I asked them if they had ever seen a Bitika or one of the other night lemurs. No, of course not. (They were little, aged about 6–8—never would be in the forest at night.) I told them about its scientific name: the discoverer named it for his charismatic professor, Madame Berthe Rakotosamimanana—the littlest lemur is named for a teacher![13]

The point is to have upbeat pictures in this film. That teacher and that little girl in the front row have enough hope for the future to lift the whole school on a hot air balloon.

The children went out in the hot hot sun and did the dances which they had prepared for us. A straightforward song about how they love nature and will save the nature of

12. Posters provided by the McCrae Conservation and Education Fund of the ecosystems for each of the Ako Series lemur books. Artist Janet Robinson.

13. Rasoloarison, R. M., S. M. Goodman, et al. (2000). 'Taxonomic revision of mouse lemurs *(Microcebus)* in the western portions of Madagascar.' *Int. J. Primatol.* 21: 963-1020.

Analabe, followed by a traditional dance with all the little girls shaking their shoulders and switching their bottoms, and then one they'd made up about fosa behavior, including little boys kneeling and batting at each other with their hands for the mating fights.

We thanked them—Emma handed out BBC pens and local sweets all round, and a small but in local terms very welcome monetary gift for the teachers and school. Then she went off to interview the village leader, M. Étienne Gilbert—the same man who was local president way back in 1985. He told the camera grand traditional tales about extinct giant lemurs. Emma speculates that by now every child in the village has some of M. Gilbert's DNA. And of course she got the consent forms signed.

Oct. 10, late morning.

David and I were left on chairs talking about the impact of children's books. What a privilege! So wise, modest, funny. David volunteered his own earliest loves: Ernest Thompson Seton's *Raggylug the Rabbit*. 'When Raggylug's mother died, I cried my little eyes out.' And *Lobo, King of the Currumpaw.* David eventually made a program on *Lobo*: they found the great sheep-killing wolf's actual skull with the bullet hole between the eyes where Seton finally, and admiringly, shot him.

Rikki-Tikki-Tavi. I asked, not so much Mowgli? 'Oh, yes, Mowgli, but the *Jungle Books* were all working up to Rikki Tikki Tavi.'

Alain Quartermain—and Umslopogas! And Blauwildebeestefontein. (I did point out that the latter was in Prester John.)

I said that for me one of the first books was Verne's *Twenty Thousand Leagues Under the Sea*. I never dreamed of identifying with Captain Nemo: it was always the Professor who knew all the wonderful names of wonderful fish. (Professor Arronax, though I didn't say it.) He instantly picked up: 'What about Professor Challenger of *The Lost World*?' I said, 'That came later.'

When David outgrew *Raggylug* and *Lobo* for animal books he went straight into the real thing: the diaries of Alfred Russell Wallace. Of course he still loves Southeast Asia.

His recall of words: to instantly bring up the names of Umslopogas and Blauwildebeestefontein and Professor Challenger is phenomenal. He is of course a wordsmith—but those are magic words, with him forever. If that doesn't show how important early stories are!!! (He is 84.)

I suddenly wonder if that eager little girl in the front row of the filming will remember the name of Bitika when she is very very old. I hope so.

Oct 10, Evening. Morondava Town.

I do not like Kirindy forest. Partly that is where so many kids were dying of diarrhea in 1985. Partly it's the discouragement of so many failed projects. But partly it's the dry, dry air—twigs crackling dry, lips cracking, the big trees gone and the jobs gone and the *tavy* fires licking the baobab trunks and the end-of-dry-season tension that feels the rains may never come back. How can eight species of lemurs and a whole raft of dedicated German students stand living here?

But 50 kilometers down the road to Morondava there is balmy sea air through coconut palms. Perfect beach hotel with elegant local-style cabins and Western-style bathrooms and king-sized beds covered with king-sized clean mosquito nets. Mangoustan rum, fresh-caught crab, shrimp *en brochette*. David restored enough to say he looked forward to tomorrow's piece to camera, to be able to pose with the sea wind blowing back his hair, chin held slightly jutting forward, gaze on the far horizon. Earlier he gave a devilish description of brother Richard always acting—either paterfamilias, or university chancellor, or even concerned older brother. So who is the greater actor?

In the end my school story was cut, though it went up on the BBC website. Sally's film, *David Attenborough and the Giant Egg*, does show Melanie releasing a tiny pop-eyed, fuzzy Madame Berthe's mouse lemur. This animal is an old hand: about the twentieth time he has been briefly caught and released; the trap may be one of his preferred food sites. (One of Melanie's interests is personality traits in mouse lemurs: this one sounds like a character.) She tells David that the 30-gram animal, small enough to sit on your thumb, can run up to 5 kilometers in a night! Melanie had tried to get the message across about the highly localized lianas and their flower bugs to several other film crews who didn't want complicated ideas, so David's piece to camera made her very happy indeed![14]

And in the film, *Madagascar, Land of Heat and Dust*, Emma Napper chose for her final story Mia-Lana Lühr's fosas. The wild ones in the forest at night were splendidly sinister. They isolated the film crew in total darkness by biting through the power cable! At last, though, the Mia-Lana's puma-like carnivores consented to emerge into daylight and full filming when a female chose her traditional mating tree to await her jostling suitors.[15]

---

14. M. Dammhahn and L. Almeling, 'Is risk taking during foraging a personality trait? A field test for cross-context consistency in boldness (2012); Melanie Dammhahn, email to Alison Jolly, January 9, 2014; S. Thomson, *David Attenborough and the Giant Egg* (2011).
15. M. Summerhill, *Madagascar, Island of Heat and Dust* (2011).

FIVE

Eleanor and the aye-ayes

Village life sometimes does look idyllic, particularly if you don't ask too many questions but concentrate on wildlife. For another *National Geographic* article I absolutely had to see an aye-aye, the strangest of lemurs, in the wild.[1] Jean-Jacques Petter had pronounced aye-ayes almost extinct twenty years before, but he announced that a few survived in the beach-front forests of the East Coast. We have since learned that they live in both rainforest and dry forest, but they are very rare and very rightly wary of humans. Everywhere there are taboos for or against. In many places they must be killed on sight. I know of one aye-aye that was killed on the East Coast and wrapped in funeral cloths. The desiccating corpse passed from village to village all across the south until it finally reached a village with the courage to bury it. Instead, in the village of Ivontaka, the taboo protected these wonderful creatures.

Ivontaka was due to become part of the Biosphere Reserve of Mananara-Nord, established by Unesco in 1987 under the direction of mammalogist Roland Albignac. Biosphere reserves were originally imagined as concentric circles: an inner absolutely protected core, surrounded by areas where sparing

1. A. Jolly, 'Madagascar lemurs: on the edge of survival' (1988).

use of the habitat is permitted, and then further outside, full agricultural use, or fishing in a marine zone, with interventions to help the population manage for greater productivity. The biosphere idea became the model for the later ICDPs, or Integrated Conservation and Development Projects: light use around the intact core.[2]

Of course this idea was threatening or even incomprehensible to people who had used the whole area before. People in Madagascar feel a profound, indeed sacred, link to their land, mediated through their ancestors, their ancestral tombs, and their immensely complicated interlace of local taboos. This means that the forest and land are theirs to use. They do not see themselves as forest destroyers, but as enrichers of the gift of the ancestors. Sandra Evers writes: 'Land anchors not only the individual, but also entire kinship groups to their history, for it connects the living with the dead… It thus lies at the heart of Malagasy cultural life. Malagasy people revere the ancestors, whom they believe to be possessed of powers that they use to exert influence over the lives of the living. Consequently, the Malagasy take considerable pains to remain on good terms with the ancestors.'[3]

The Western view of land as mere property which can be bought or sold is in fundamental conflict with deeply rooted Malagasy values. This conflict will reappear throughout this book: in attempts to 'rationalize' land tenure, in foreigners as land-grabbers in local eyes whether investors or conservationists, and even playing a part in national government change. But back in 1987 people of Mananara town and Ivontaka

2. Unesco Biosphere Reserve Information: Madagascar, Mananara-Nord, Unesco; R. Albignac, G.S. Ramangason et al., *Éco-développement des communautés rurales pour la conservation de la biodiversité* (1992); G.M. Sodikoff, *Forest and Labor in Madagascar: From Colonial Concession to Global Biosphere* (2012).

3. S.J.T.M. Evers, C. Campbell et al., 'Land competition and human-environment relations in Madagascar' (2013), p. 9.

village just knew the Biosphere Reserve was coming, even though as yet it hardly touched their customary ways.

July 1987, Mananara and Ivontaka. Letter to Mother.

Dear Mom,

I shall tell what I've been up to and please don't be put out with me. Remember you let Pop read me all Rider Haggard, and you weren't much more stay-at-home yourself, what with your *St. Nicholas Magazine* firing you up to come to Madagascar and find an elephant bird.

I flew to Mananara last Sunday with some trepidation to join Eleanor Sterling. I had a list of names from Jean-Jacques Petter and Roland Albignac, and a mission—this region is to become a Biosphere Reserve for the aye-aye. I also was fascinated by aye-ayes since my Bastille Day visit to Vincennes Zoo. They feel everything, like raccoons, not at all like a lemur. Their minds must be as different as their looks. Humphrey, the adolescent male in Paris, ran round poking his finger into everything like a naughty child in a supermarket.

I lumbered off the Twin Otter with my camping duffel. There was Eleanor Sterling, ever so pretty in washed blond hair, glaucous hazel eyes, and a purple full skirt she'd 'borrowed' from her mother. It turned out she was recovering on Sunday from a grueling week running up and down hill 20 or 30 kilometers a day, with guides who were really hunters that swarmed up trees to shake and poke every aye-aye's nest they found. Eleanor was captain of her cross-country team at Yale: the background that's needed for the thesis she's planning. She has given up her job with the World Wildlife Fund and is now a grad student of Alison Richard's at Yale. She is all prepared to gallop after aye-ayes all night long. That is, if she can find any.

Awful day yesterday when she'd been in one place and they actually found an aye-aye in another. Over her protests she was confronted by an animal with a noose round its neck, trussed up in a gunny sack, and five guides who demanded a $30 reward from her meager student funds because Frans Lanting the photographer bought aye-ayes for $50. They proposed to take it over the river in a pirogue to an island where Roland Albignac wants aye-ayes released. They actually got there before she paid the $30. She insisted that now it was her beast so they were taking it back where they found it, and she was coming along to make sure it was freed and not stewed. So they did half of it by river, the pirogue loaded to the gunnels (such as gunnels are in a dug-out) and the rest on foot, Eleanor shaking with misery for the poor trapped animal and fury that it was so hard to impose or even explain what a field study should be.

So when I turned up she knew the landscape and the guides, but also knew she couldn't go out to watch at night without a car, a grant and a mastery of Malagasy—in short, try again next year. Meanwhile, every girl in town was her friend. She chaperoned them to the local disco. One youth had taken her touring to the southern coastal villages on a motorbike as preparation to a proposal of marriage on behalf of his older brother. Eleanor will go far. She even got me to the disco, which turned out to have flashing ultra-violet lights and an American rock tape singing 'What will I do for a role model, now my role model is gone!' You can imagine white eyes and teeth and dresses jiving in the ultraviolet lights. This in a building of palm-thatch and hand-hewn planks, with a glittering driveway of crushed quartz crystals. The rosewood-colored youth of Mananara do sedate jitterbug in their disco, not the lascivious shimmy of the villages.

Monday we tried again near Mananara. We walked for 6 or 8 hours unsuccessfully. This is all farms and villages—rice fields, coconuts, bananas, vanilla pods drying and the sweet pungency of coffee flowers. There were tufts of woods in between with a few bunched stick nests like eagle nests among the lianas, without any aye-ayes inside. I got carried away at the end when an old man, a vanilla grader, said an aye-aye lived just behind his house 2 km off. Of course it was twice as far, but I figured our only hope of seeing an animal was if we got to a nest at dusk. No hope at all. Still, it was a fabulous walk in the sunset, with rows of hills in artful views, and coming back in the dark through villages with the evening fires lit, and the Milky Way blazing across the sky.

Unfortunately we wound up on the wrong side of the Mananara river. Our guides had to liberate a locked-up pirogue fastened to a post by a massive chain. The lock was a nut and screw. Only the pirogue's owner was rich enough to own two pairs of pliers to unlock it. He took us across—maybe 75 yards of still warm black river, with me kneeling and holding my breath the whole way, given only two inches of freeboard. We landed just in time, when I was deciding I'd be no good under torture and would have to stand up, tipping out all the people and the cameras whether or not there were crocodiles in the river. In the village on the other side a pick-up truck stopped as we flashed our lights and halloo-ed so we rode home the last 5 miles in the warm wind under all those stars, while a voice among the other passengers explained he was a teacher of mathematics from Rouen, home visiting on holiday—explaining in English!

But Eleanor said on her motorbike ride last week she had seen a most beautiful village on an idyllic cove with real rainforest behind. Even before that exhausting day's walk in the farmland of Mananara, we arranged with merchant Nim

Tack to board his Mercedes and ride south to Ivontaka village.

The 18 km to Ivontaka is mentally further than New York to Tana's Hotel Colbert. Only a Mercedes can make it—a dinosaur with about 4 foot clearance. Its driver is King, like Mark Twain's Mississippi steamboat pilots. He is sensitive to every incipient skid on the laterite, and strong when rocks yank the wheel, and cool in choosing his course through the ruts and ravines. At times he orders the passengers out—Eleanor and me from the cab, 31 more from the back, not counting babies. His work gang unloads planks from the undercarriage to reinforce bridges. An assistant runs before like John the Baptist for about half the way, and guides the wheels onto the bridge-planks with upraised finger. John the Baptist barefoot, of course, in remnants of shorts and a T-shirt consisting mainly of its neckband. (Of course there are rumors that the Mananara merchants do not want the road repaired, so that their boats and massive trucks like ours can gather the cash crops of vanilla and cloves with no competition from less well vehicled interlopers.)

In every village that we pass, the prettiest girl runs up with a folded note, for the driver? Or for a swain in the next village down the coast? Or she simply looks sideways at the driver from under her curly eyelashes.

It took four hours to drive 18 kilometers—and we hadn't even been stuck.

We (that is, three guides we brought from Mananara, E. and me) arrived at last at Ivontaka. As Eleanor promised, it was perfect. The coast bent into a little cove guarded by granite points and a little fringing reef and lagoon. Perhaps thirty huts of coconut thatch, and a double row of coco palms next to the sea.

Right, Mom, Floradora scene setting all over again. The chief and owner of the four-cup café, M. Louis Rabeson

Rabearivelo greeted us in most elegant French. He only moved back from Mananara four years ago. He invited us to lunch with his family. His home is one room, with benches six inches high along the walls to sit on or store things, and mats unrolled in the center. Young wife in orange and ochre *lamba* and lime green shirt, and two poppets of kids. It may be a village, but our host told us stuff about trade winds and pluviosity and asked if we had come to talk about the future Biosphere Reserve. He also told us an amazing tale of the village of Soanierana Ivongo, where the *tavy* cutting has so deforested the hills that their river has dried up.

We gave his small daughter a meter of pink flagging tape for her braids—by morning she had shared it with every girl in the village, so each had a butterfly of fluorescent pink in her hair. The chief assigned a leprechaun-like local in pink sateen shorts to take us up the forest mountain behind, and fed us and our original three guides a mountain of rice, with thin bouillon of a watercress-like green, and a very few tiny dried fish, mostly salt. Then we started uphill to camp. We had to stay out of what looked like a nearby, promising bit of rainforest—that part is *fady*, taboo—perhaps it holds tombs.

At this point I discovered I had walked too far the day before. I could barely climb, and soon took to lagging behind the rest. It was nice forest, but who cared. After an hour we reached the summit and put up our tents—a little one for Eleanor and me and a big one for the guides. Eleanor wanted to go on because the guide said there were black-and-white ruffed lemurs in the woods. We heard them roar in the distance. The leprechaun suddenly started a shouting sing-song, which turned out to be a prayer that we would see them. He and Eleanor and the others set off—and the prayer worked, which I knew by the bellows and squeals that ruffs give when alarmed.

It poured all night, and neither tent leaked, which is a lovely smug sensation.

Everyone agreed there are no aye-aye in primary forest. Eleanor and I aren't convinced, but at least we'd seen what the forest looked like. So we all slid downhill, which had somehow got much lower since I climbed up it. The others pointed to another vertical slope of rocks and nasty scrub. I let them and Eleanor tackle it while I snoozed beside, not under, a coconut palm by the sea.

By the time they skidded back all bruised and scratched and muddy, even the guides were tired. It's not often a Malagasy guide admits that.

They took us to one last place—a coconut grove well beyond the village where many of the nuts had the round gnawed holes of an aye-aye's chisel teeth. There was a nice stick nest with fresh green leaf-lining in a tree beside the biggest palm.

Eleanor and I opted to camp in the grove that night. We sent our guides back to town for lunch and suddenly realized we'd found the only deserted beach I've ever seen in Madagascar. We pulled off all our filthy clothes and rushed into the turquoise water and bobbed up and down ... and then rushed back up and dressed, sure we'd horrify some villager—but there was nothing to do but sunbathe all afternoon on that lovely shore. Eleanor with her Presbyterian conscience tried to feel guilty but I don't think she did.

The guides came back at 4.00 with our gear. We camped with a tent each on the soft coastal beach grass and we went and sat under the nest, just Eleanor and me. In the very last light a ripply shape ran along the branch, and up a coconut palm frond. My headlamp picked up orange eyes spaced wider apart than the heads of two normal lemurs. Eleanor turned on the spotlight, and there was our aye-aye, classically chiseling open a coconut.

I didn't think it would be handsome. That hefty head, and bat ears, and the hands like bunches of knobbed licorice sticks doesn't seem as though it could even come together to make one animal. But it undulates as it moves with that great plume of tail rippling behind. The palm fronds did not move at all under its feet: it's like a hologram skimming above the frond.[4] The face has kohl-rimmed eyes, set off by the triangles of pale skin above and below. At least it is *jolie laide*. And it was curious about us, too—it hung by its hind legs, arms hanging down and ears spread, to goggle at our lights.

We watched for about 40 minutes, hiccupping with excitement while it scraped the nut meat out with flickering finger just as it's supposed to. But then our faithful guides, all four of them, came back from their supper and clearly a tot or two of village-style sugar-cane beer. Four sets of crashing feet were too much. The aye-aye fled, fluently, for the hill forest. I got to the guides and told them they were nice to come back in the night to help, but PLEASE GO AWAY —then fell full length on a dry bush, which settled the matter. The guides did go, begging for my headlamp to see their way home.

Eleanor and I roamed hopefully through the groves, but no more aye-aye. Instead we found that the hut-sized rock which overlooked the lagoon, where I'd dislodged a very smelly squid to sit and watch that afternoon. It now had a squid put back, and, as well, a scarlet-and-mustard colored armless starfish, a pentagon big as a man's hand. The rock was clearly fady, taboo, with its offerings from the sea. I had all too clearly violated the taboo. Eleanor reads too much anthropology. I mischievously remarked that nocturnal

4. An engineer told me later that this must mean active damping out of the normal waves caused by locomotion. In other words, a way to creep up on sleeping birds or lizards, and a very good way to avoid notice by their own predators—human beings.

primate-watchers are occasionally taken for *pakafo*, the dread white eaters of children's hearts. This is, on rare occasion, very bad for the scientists.

'Look! Eyes!' Down through the bushes by the sea gleamed orange sparks large as aye-aye eyes. As we approached, they reflected from still water. Men walked apparently on the sea surface, holding up flares of burning splints that gleamed off their brown bodies like Goya night scenes. They were walking the reef at low tide, fishing with long spears in the glassily still black lagoon, while the surf crashed impotently just beyond.

Another group of flares progressed toward us from down the beach. 'Please put out your light. Please don't be looking through your binoculars as if we're spying' quavered Eleanor.

'But look,' I said, 'They've a bunch of little boys with them. Just walk up and say hello.'

'But they think—that word I won't say—come to eat children.'

At this point the group arrived. The kids ran on giggling, but the men stopped—our guides. They'd been using my precious headlamp to fish, and delightedly held up three cornucopias of banana leaf filled with tiny pale pink crabs. The lobsters on this coast are cobalt blue with purple tails and the crabs are peach-blossom pink.

They'd found us an aye-aye in its nest—let them eat all the crab they please![5]

5. Eleanor later studied aye-ayes for her Ph.D. for two years on the uninhabited island of Nosy Mangabe. She has become director of the Center for Biodiversity and Conservation at the American Museum of Natural History, with work around the world, especially in Vietnam. See E.J. Sterling and E.E. McCreless, 'Adaptations in the aye-aye: a review' (2006).

PART II

Politics

From 1985 to 1991, Western conservationists saw a huge opportunity: a political moment when it might be possible to mobilize funds and energy to save the extraordinary natural riches of Madagascar. The environmental movement had taken off in America. Now, here was a semi-virgin country: a country of socialist xenophobia which then announced in its 1985 conference that it wanted aid for its environment. No wonder we jumped in.

The next chapters range from the mid-altitude rainforests of Andasibe to the meeting rooms of the World Bank and back again to Antananarivo, the country's capital, as the Bank and foreign donors and the Malagasy government argued out their aid package. Even the donors were infected by idealism for a project to tackle the most urgent environmental needs of an entire country. The Malagasy, long battered by colonialism, were more sceptical of this new effort to foist large loans on the country in the service of a foreign ideology.

By 1991, the package was triumphantly launched as the NEAP: The National Environmental Action Plan. The NEAP's multifaceted project had its own secretariat to coordinate multi-donor funding, and a prospective scope of three stages over the next fifteen years. The six-year journey from 1985 to 1991 involved a huge amount of effort, idealism and controversy.

Chapter 6 tells of the 'seed money' pledged by the Alton Jones Foundation in 1986. The part I tell of their ambitious tour is only in the rainforest of Andasibe. I couldn't resist adding a vignette of my own first view of lemurs in that same forest back in 1962. In 1987, Chapter 7, that seed money supported the comedy of a high-level Malagasy delegation touring American zoos. The Malagasy did gradually begin to understand where all these foreign lemur-lovers were coming from! Chapter 8, 1989, is the World Bank in Washington, and Chapter 9, 1989, the donors in action in the Antananarivo Hilton and back in Andasibe, thrashing out the NEAP. Chapter 10 is the tragedy of teenage Bedo, an allegory of the perils of too much money in too poor a land. Still in 1989, this period culminated in the apparent triumph of conservation planning for Madagascar. In Chapter 11, Barber Conable, president of the World Bank, drags much of the Malagasy cabinet forth to hear the indri sing in the rainforest of Andasibe.

SIX

Where indri sing

In my very first weekend in Madagascar, way back in 1963, I escaped from town. My goal was the village of Perinet in the nearest bit of rainforest. Preston Boggess, a Yale senior whom my professor, John Buettner-Janusch, designated as my assistant, and I were both wild to try out our brand-new Land Rover. We gleefully drove eastward along the road that hiccupped up and down over the bumpy plateau hills, and then flung itself in wriggly curves beside a cascading river in parallel with the equally wriggling railroad on the river's opposite bank.

Perinet was named for the French engineer who designed the railroad line between Antananarivo and the main port, Tamatave, now called Toamasina. Nowadays Perinet is called Andasibe, which means the 'big camp.' This memorializes not the French engineer, but the workers who camped in the forest to build the railway and founded a village near the line.

In 1963 there were still daily up and down trains, which crossed at lunchtime on Andasibe's double section of tracks. The Buffet de la Gare awaited the two drivers and their first-class passengers. Both trains stopped for an hour and a half, the time allotted for a proper French midday meal. Second-class passengers could go off to buy rice and stew at local *hotely* or gobble bananas and dried fish and fried bread balls from train-side vendors.

The glorious song of Andasibe's indri comes back often in this story. The Hotel de la Gare with its pillars and its floor of ancient rosewood also reappears—I forget if it was Gerald Durrell or the Duke of Edinburgh or the president of the World Bank who found himself sharing its best bedroom with a pigeon flying in through the broken windows. No, the Duke of Edinburgh did not come. So it was Conable of the World Bank. But that was later. When Preston and I first turned up the hotel maintained its French cachet: multi-course meals offered on a stacked-up series of plates, and waiters who wore white gloves to serve us although their feet were bare.

The rainforest was very different from what I'd imagined. At mid-altitude on the escarpment there was no solemn majesty of lowland forest. Instead a plethora of spindly interlaced white-trunked trees opened out to glimpses of bright blue sky and distant hills. Among the trees bounded the first lemurs I ever saw in the wild: gray bamboo lemurs, a grunt-grunt-grunting troop of brown lemurs, and indri! Largest of living lemurs, black-and-white indri sing the song of the forest as whales sing the song of the sea. I confess they made me giggle because when they lifted their heads to roar at us in alarm, their mouths pushed out into curved red lips like 1920s' flappers. I almost expected their bellowing to end up with 'Boop-boop-be-doop.'

Just as we left on the second day came the truly breathtaking moment. A storm cloud loomed blue-black behind the forest. Then a troop of diademed sifaka leaped above the path. Diadems may be the most beautiful of all the lemur tribe: black faces rimmed in white ruffs, charcoal gray backs, and golden-orange arms and legs. Late sunlight spotlit the gold and white and warm gray animals soaring above the trees. Five of them, one after another, in single file, the arc of their incredible ballet-leap repeated, repeated, with their gleaming orange limbs and white-rimmed heads before the lowering blue-black cloud—and then they were gone.

And then they were gone. For many years afterward no diademed sifaka survived in that forest patch. Indri were taboo to hunt, untouchable, but sifaka in the decades of lax guardianship became merely meat. Today's tourists may see them again, either in the great adjacent National Park of Mantadia or in that same little forest saved for the indri, where a few groups of diademed sifaka have been reintroduced to their ancestral home.

I have been back many, many times after that first glimpse—in 1986 with the Alton-Jones family and their friends.

Tom Lovejoy, whom I had known since Yale, and Russell Mittermeier, energetic enthusiast for primates and reptiles, asked the W. Alton Jones Foundation to fund preliminary work in Madagascar toward a conservation initiative. This could become seed money that led on to major funding by international donors. Tom is the man who began the great project in the Brazilian Amazon to scientifically document and analyze the impact of chopping forest into small fragmented tracts—a project which has grown in importance over almost five decades.[1] As for Russ, I wrote:

November 1985.

Russ was described last year by the Minister of Eaux et Forêts as 'that man who looks like a movie director.' Russ hails from the Bronx via Harvard to the Amazon, has long curly hair, a face maturing from simply handsome to weather-beaten smile-lines, and a penchant for collecting blowguns. He was 38 last Saturday and will probably keep running through jungles like a 20-year-old when his hair is long white curls. He

1. For a recent appreciation of Tom's influence, see J. Tollefson, 'Splinters of the Amazon' (2013).

administers the galloping World Wildlife primate program by telephone from Stony Brook, where he leads a largely nocturnal existence so he can get his writing done undisturbed. Russ, incidentally, speaks five or six languages... English, Spanish, Portuguese, French, his maternal German, and Sranan-tongo, the Creole of Suriname.

Tom and Russ were not modest. They wanted $500,000 for a country off Africa which most people at that time had never heard of.[2]

The Alton Jones family decided they should investigate Madagascar for themselves. Tom's personal contacts through Millbrook School and Yale University convinced them that they would be in good company. Several of the family signed on: Edgertons, Johnsons, and Aunt Betty Jones, as well as their fund manager Jeff Keleher and his wife Louise. This already glittering assemblage then invited their friend Olga Hirshhorn, who had recently donated a whole Museum of Modern Art to the Smithsonian Institution. Also Scott and Hella McVay, whose family foundation supports educational initiatives in New Jersey. And Carleton Swift, ex of the CIA, and Henry Mitchell, columnist for the *Washington Post*. At the last moment Dillon Ripley and his wife Mary decided to come too. Dillon had just retired as secretary of the Smithsonian Institution—we were now twenty-four years on from when he'd been on my thesis committee at Yale, and inspired the undergraduate ornithologist Tom Lovejoy.

Why was I asked along? I never have held an official position, even in the universities where I taught. I guess by 1989

2. Both were at that stage with the World Wildlife Fund. Russ later helped to found Conservation International, where he has been chief visionary and scientific advisor. Tom became president of the Heinz Center for Science, Economics and the Environment.

I was a fixture, a grand old lady aged 52, and an American who already had kicked around in Madagascar for twenty-five years. Also a presentable friend of Tom and Russ.

Now, funding for all the next steps in Madagascar's conservation hung on whether the assembled millionaires were enchanted or repulsed by Madagascar.

November 2–3, 1987: Perinet. Settling in the millionaires.

Up at 5 a.m. for the train to Perinet. We follow the waterfalled ravine. The train almost loops the loop to wind down the escarpment. Bobbie Johnson counts seventeen stops to our destination. Lovely, gentle ride.

Four separate groups are trying to squeeze into the Perinet Hotel de la Gare. Our eighteen millionaires, plus Roland Albignac squiring ten Italian TV men, plus botanist Armand Rakotozafy with three palm people from Kew Gardens, plus a trio of bug-catchers who seem to move around on their own. The palm people—John Dransfield and David Cooke of Kew, Jim Beach of Missouri, and Armand Rakotozafy of Tsimbazaza—are in one of those paroxysms of enthusiasm that afflict biologists here. They've been hiking just for one day here and come up with a palm that hasn't been seen for sixty years, and even then was known just from a single specimen. On top of all, Barthélémy Vaohita, the Madagascar WWF rep., has stirred up so much local gusto there is now an Association pour la Sauvegard de la Nature. Our hotelkeeper, whose business depends on nature-watchers, is president, proudly wearing one of Russ Mittermeier's lemur T-shirts.

The housing crisis is solved by banishing Roland's group to Moramanga, an hour up the road, Armand's to the 'annex' which Tom has already spurned, the bug-hunters to the rat-infested Eaux et Forêts guesthouse, and putting all four single men—Carleton, Henry, Russ and Tom in

'Millbrook dorm,' the big room at the hotel's end upstairs. This leads to trouble. Henry Mitchell is locked into the dorm room by Carleton. Henry puts his head out of the window, shouting 'Au secours! Je suis prisonnier!' to the populace of Perinet. I bet he read the *Prisoner of Zenda* aged 6 and never outgrew it.

Frans Lanting, who passed through before, briefed 17-year-old Josef Bedo, teenage guide extraordinary, to show us around. Even with his help we don't find much in the woods except for Dillon Ripley, who is hypnotized by an asity with an emerald eye. But Bedo tells me that there are aye-aye in Perinet, and indeed that they may appear in twos and threes! No one else knows this—they are so rare they have never been recorded in these rainforests.[3]

November 4.

How I hope the indri sing! I've been dreading this morning since Scott McVay read me back a poetic paragraph about indri from my 1980 book *A World Like Our Own*. I can't get away with this when he does write poetry, and his family foundation sponsors real poets and poetry festivals in New Jersey. Or even if I can, the indri just have to sing to prove it. Quite aside from his wanting to compare them with his studies of whales.

But little Bedo and his forestry gang have been out since dawn—they'll at least find the creatures.

Of course they do. Distant singing as we approach—then the indri, languid and lordly, recline above us in their trees, while Russ Mittermeier manhandles his tripod on the steep slope with the telephoto lens he probably filched from Palomar Observatory.

This is group A, a group of six, Jonathan Pollock's original habituated friends from the 1970s. They glance down

3. Velvet Asity, *Philepitta castanea*, one of four species in an endemic bird family.

over their long black muzzles. They do not leap with their 2- to 8-meter ballet jumps. This lot are not to be budged an inch by eighteen tourists plus ten Italian film men chattering underneath them. Roland Albignac and I earnestly discuss the future of the new Mananara reserve for aye-ayes. This may be the best chance Roland and I get for a talk.

The Italian TV presenter wears a chronograph watch you could use to fly a Boeing. It nestles among the blond hairs on his tanned and muscular arm. Tired of waiting for the song, he does a piece to camera while indri loll in the background. His program is called something like *The Adventures of Jonathan*. His cameraman is busy making him look intrepid among the indri, aided by the odd boa constrictor and leech. For intrepid, give me Betty Alton Jones any day, stumping along through the rainforest with her cane and her duck-headed umbrella. Or Carleton, who doesn't complain about hollow beds and a bad back. Or Mary Ripley hefting that demolition bomb of a camera case. Or the Italian cameraman himself, who got a leech off the camera under his eyelid and is still shuddering.

The camera crew manage to shush up the assembled indri-watchers so the muscular blond Italian can begin his piece to camera, 'Here I am all alone in the wild jungles of Madagascar...'

After we've waited perhaps two hours while the indri take up various Madame Récamier poses, they finally sing. The B group begins in the distance; I switch on the recorder. Ours lift their heads in the barking roar of introduction, then the full song, the four notes of descending wails, each note a third (almost) from the one before. It's a duet: the female sings the first three notes and the male the last, lowest note of each arpeggio. They pause while group B replies antiphonally. Ours take up the song, now with variants, three or four voices overlapping. B replies again, briefly, and ours close with one low moan.

No one can say much...

Except Scott, later. He just wants to borrow the tape. He sits with headphones, eyes closed, in the funny hotel dining room at Perinet, composing a poem of his own.

Tonight we go out with Bedo for a little night ramble...

How clever of Tom to spot lemur eyes all on his own in the moonless night among the interlaced forest trees! He and I eventually find his mouse lemur. Oh dear, we then find we have lost the other party. Abandoned in the rain-forests of Madagascar! Adventures of Jonathan, or rather of Tom Lovejoy! We rather cleverly steered out of the woods towards Orion all by ourselves...

Pat Edgerton, Betty, Carleton and I sit and talk. Pat outlines her hope that Madagascar will be a model for other nations. A place that is an island, a continent and a country, blessed with a treasure-house of evolution, should command enough foreign attention to do anything at all with its environment that it chooses to do. And when 80 percent of its people are rural, living from the crops they grow, the water they carry, and the fuelwood they gather, they themselves have every reason to save trees, water and soil. It's all so clear in the long run. But then it ought to be clear the developed world is insane to continue the nuclear arms race. People do cling to insanity.

I cling to the thought—no, the fact—that people of vision do change the course of history. Things are changing within Madagascar, too. This is the moment, the first in years, when progress is possible... with the help of the Alton Jones Foundation.

SEVEN

Napoleon versus the zoos

The W. Alton Jones Foundation stumped up the 'seed money' for Madagascar, but much more would be needed. Russ Mittermeier decided that Malagasy politicians and civil servants should see for themselves how much Americans value endangered species. Only then, if the Malagasy were convinced, could they move on to request big bucks for conservation from the World Bank and the aid donors.

Thus the intermediate step was a tour of American zoos for a select group from all concerned Malagasy ministries. The group began by brokering an agreement on the importation and ownership of lemurs to make the zoos happy, and then toured a few places which breed endangered species.

We met in May 1987 to discuss the zoo agreement in the unlikely setting of a barrier island off Georgia. The place has a history: Spanish missionaries, epidemic deaths of Native Americans, a plantation house that once belonged to Button Gwinett, who signed the Declaration of Independence; also cages belonging to the Wildlife Conservation Society, with endangered macaws and macaques and a breeding pen of the world's rarest tortoise, the Madagascar plowshare. Free-ranging ring-tailed lemurs sauntered up to the visiting Malagasy, tails in the air like swaying black-and-white question marks. Minister Randrianasolo pointed and announced, 'It's a protest demonstration!'

The human mix included Joseph Randrianasolo, Minister of Water and Forests; Henri Rasolondraibe, secretary general of the Ministry of Scientific Research; Mme Berthe Rakotosamimanana of higher education; Voara Randrianasolo, the director of Tsimbazaza Zoo (no relation to the minister); and Barthélémy Vaohita of WWF Madagascar. The only one from the minister's own ministry was Joelina Ratsirarson, who wound up carrying the minister's handbag-purse, and his briefcase, and his stacked-up armfuls of presents given by the zoos. It amused the minister to treat him as a bag carrier. It also amused the minister to browbeat all the other ministries into submission, not to mention the assembled Americans. I'll tell the story, but first here is the summary I wrote to our daughter Susie.

June 1987, Cambridge. Letter to daughter Susie.

I haven't written in a month because I have been travelling with the mad delegation from Madagascar. I had all these plans to write a book about their visit—you know, the conflicting arguments of the Malagasy view of why lemurs are important, and the American scientists' view, with an excuse to tell some of the science.

What happened was a novel, instead. The minister who headed the mission wishes to be a cross between Napoleon and Peter the Great. He dominated men by simply outlasting and outdrinking them, women by charm and the occasional proposition if the woman seemed intelligent enough to be a challenge, and both by sheer political fast-footing. Whenever he was not around everyone talked about what he had just done or what he was likely to do next, so we were never actually out of his company. The night before we left America, in the Great St. Louis Tudor

Bedroom meeting, he staked his claim to the control of environmental policy in Madagascar with the rest of his delegation sweating fear like dogs. This is an incredible development, in that it means that a strong minister thinks that the environment and the little lemurs are a prize to be seized—a leading edge which will trail the World Bank and other interests behind it.

If, of course, he isn't assassinated first.

Then picture the décor—Spanish moss and live oaks and a Revolutionary War plantation house on St. Catherine's Island, the executive suites and embassies of Washington, the pseudo-South Pacific Hotel of San Diego on a bay full of navy warships. We went on to the Cheshire Inn of St Louis: Mock Tudor studded with maces and halberds. And on, because the momentum was such that Russ Mittermeier of WWF, who'd set up the tour, and I decided that we were not going to let this go sour at the end, to Jersey which has already done so much for Mad. Lovely Jersey Zoo where your orang-utan forced himself on his female in front of the delegation as usual. So the circus ended with red ruffed lemurs in the mossy beech woods, and a pink granite manor house, and the minister telling strange orations from Diego Suarez in a *real* Tudor pub by St. Ouen's bay.

Russ Mittermeier is no mean character himself. The minister first met him in Bezà Mahafaly at the dedication of the place as a new reserve. The minister pointed over at Russ, then attired in silver running shorts and silver singlet and brandishing a couple of Antandroy spears, and asked me, 'Who is that man who looks like a movie director?' I admire Russ immensely for setting up his caper—I don't know anyone else who could have, except Dad.

Also with us was Long John Hartley from Jersey—totally British in manner and fiber—understated and hilariously funny. And spherical Madame Berthe, whom I used to call the Tiger of Tananarive until she and Barthélémy Vaohita turned up in Jersey in 1983 and turned out to be perfectly sweet. Also Barthélémy quoting Montaigne. Also Joelina Ratsirarson, who has spent years grousing at foreigners for neglecting Water and Forests, and now may be guillotined by his minister for not having attracted foreign interest and support for their ministry. Ratsirarson is about 4 feet tall—and is also capable of being sweet when he doesn't have to be nasty. At the final end of the final dinner Long John leaned down from the stratosphere and picked up Ratsirarson in his arms, who kissed him on the cheek, and held the pose for Russ to photograph. After, of course, the minister had soared off to Paris in the plane personally chartered for him.

All these people, you realize, have known each other and fought each other from their respective ministerial bastions for twenty years. Antananarivo is somewhat smaller than Ithaca, New York. You can imagine what burying of hatchets it took to get them all on a mission together in the first place. You can also imagine that they will have to live with the results of this for at least the next two years—or maybe twenty.

When it was all over—probably the most intense three weeks of small-group psychology of my life, except the weeks just before Dad and I got married—I went up to see Margaretta.[1] She had just finished exams at Cambridge and was in a kind of euphoria that floats out of contact with any of the world below. When I came in at midnight I told her all about the minister till two in the morning. Then we talked all the way to Grantchester the next day and had tea in the orchard there. The river banks were full of undergraduates on green grass in the sun, looking like one of those unnaturally bright collages that used to come out of

1. Our older daughter.

cereal packets. The river itself was a forest of punt poles outlining the distant spire of the church, with snatches of overheard conversation: 'It was just here, at eleven-thirty at night, that Elizabeth and I fell in...'

Margaretta still hadn't emerged from the American literature bubble, so she talked about *Moby Dick* most of the way. Apparently she was so wound up in it that her personal tutor forbade her to read *Moby Dick* for two days on grounds of health. She had discovered a wonderful book of criticism about the whale as Ahab's externalized body. We decided that the minister was Ahab with the presidency of Madagascar as his whale, dragging his crew with him by sheer epic potency. Call me Ishmael.

❦

The Madagascar delegation arrived in America jet-lagged, then shell-shocked by the Bronx Zoo's multimillion-dollar artificial rainforest with its artificial thunderstorms, and then were kept awake arguing for the whole three days of the St. Catherine's meeting. The main dispute was ostensibly about the 'descendants' clause.' The Malagasy delegation wanted any newly exported lemurs to remain the property of Madagascar, as well as their descendants. In the 1980s zoos were beginning to cooperate over endangered species, with studbooks that allowed them to plan the best possible matings for conserving genes. The assembled zoo curators said they could never agree to outside control over the descendants.

The real reason was that the minister wanted to exert control—not so much over the weird Americans as over his own delegation. He finessed the discussion to put his own lot on the losing side, then magnanimously forced them to agree that the Malagasy lemur's descendants should be rationally managed according to the zoos' protocols.

Here is a little flavor of the closing night:

Thursday, May 14, 1987. St. Catherine's Island, Georgia. Somewhere in the afternoon I snuck out to visit the lemurs—Portia and Romeo with their radio collars. Exiled Kate is limping along with a slashed right thigh. Madagascar's people conferred in the front porch without their minister, who was ensconced in a wicker chair, his hand over his eyes. I was sent in twice to ask the minister about modifications to the descendants' clause.

Funny thing—three new clauses are proposed by the zoo people to bind *themselves*—no species to be exported without a management plan, all animals to be entered in the overall ISIS database and also in species' studbooks.[2] These are far more demanding of zoo time and general commitment than simple nominal ownership. Clear that we are just at turning point from private to general species management, with different sticking points. I asked Diane Brockman of Los Angeles about how zoo exchanges work; in fact ownership often becomes a burden, and a way of assigning financial responsibility for all these offspring. Ownership certainly doesn't imply enough local zoo control to contradict the species' cooperative breeding plan.

At 6, the minister took his hand from his eyes, rose and went to sort out his people. Assembled westerners cheered.

Twenty minutes later they were back. They had a further set of modifications, including softening the management plan clause to simply say only exported for breeding. These sailed through.

The min. said ownership clause would be on exported animals. Nothing to be said about descendants.

It sailed through.

2. ISIS: International Species Information System, which holds data on individuals of endangered species in captivity worldwide. www.isis.org.

Then Alison Richard of Yale, Mme Berthe of higher education, and Dan Wharton of the Central Park Zoo went off to an outbuilding, ancient slave quarters, to translate and type revised versions, while the rest of us had supper and the minister went alligator-watching in the island pond.

At 9, the translators had not reappeared so I carried them some beer and brownies remaining from supper.

Yet a further crisis—a jeep full of Voara, the sec. gen. and Ratsirarson drove down to the house were the translators were working. As they left the minister said to me, 'I sent them to make sure there was no reference to descendants.' However, as they drove the three were shouting at each other in Malagasy. When they arrived, the sec. gen. put the descendants' clause back in, standing over Alison R. as she typed, directing her in French.

'Hang on, that wasn't my recollection,' said Alison R.. 'Shouldn't we phone the Minister and check?'

Madame Berthe of all people came to the rescue. 'I have the minister's original text here,' she said. And there were the descendants crossed out.

So they all went away—and the French was typed, and xeroxed and stapled, and then they read it through and realized descendants had crept back in on another page. We fiddled and changed the English one, and unstapled the French to remove the offending page.

At 11 p.m. Alison R. turned up, saying everyone was waiting exhausted and increasingly angry in the Big House. We kept on correcting.

Then Ulysses Seal turned up and shouted that he could have typed the whole thing in twenty minutes, so he was going to take charge.

We went on correcting.

At 11.30 we jumped into the jeeps, swinging on the roll bars, arms full of xeroxed copies. I shouted 'A-team to the rescue!' John Hartley gunned the accelerator. We bumped 200 yards over the compacted sand, while white-tailed deer bolted in all directions, their eyes rolling and gleaming in the oncoming headlights. Then, of course, we stopped since cars can't cross the bridge from the reconstructed slave cabins (or the ruins of the old ones—forlorn fireplace blocks of *tabbe*—cement with shells instead of pebbles).

We pounded into the Great Hall, where tired bodies slumped in chairs in the shadows. They twitched and moaned and growled at us, like casualties of some soporific gas. We handed round copies, which they made a show of reading through.

The four copies were laid out on the table—which we all signed, rather than initialing, the minister last, with the lamplight from the peony-painted antique Chinese lamp on his face, sideways against the dark vaultings of the hall.

So, still on the night of Thursday, May 14, we stood saying goodnight at midnight, with a vague sense we should be celebrating. John Hartley and Roland Albignac made separate queries where there was a bottle of champagne, standing under the musicians' gallery of the Great Hall in the part of the house which belonged to the signer of the Declaration of Independence. I pointed out we'd just signed a Declaration of Non-Independence. The Malagasy delegation walked past in that peculiar stiff-legged fashion of those who've been awake for forty hours and are promised sleep. I went up to the ancient North Bedroom, which I share with Mme Berthe, where she was already KO'd and beginning her almost narrative style of gurgling snore. That house smells like my mother's: old, old wood, and lavender and mothballs.

A tap on the door—would I join Diane Brockman and Alison R. and John Hartley next door? Berthe was invited, but she declined.

Diane undid a plastic hip flask—quart size, not pint—of Scotch. I fetched my toothbrush glass. We all toasted happy conclusion. Alison R. said she was going down to see the minister—we said, 'If you're not back in seven minutes, do we mount a rescue party?' She claimed she could take care of herself so long as she could say we were expecting her upstairs.

But she was back in five to invite us to join them.

So we came down to the grand bedroom with its double high-posted beds, and its portrait of the Declaration signer, its amazing lavender bath suite, and a bouquet of magnolias the size of dinner plates. We still clutched our toothbrush glasses and held them up like Oliver Twist for a tot of the minister's Chivas Regal.

We talked and talked but it would be hard to remember much.

At 3.30 somehow we all decided to go to bed.

What have we done? Exalted our small-group psychology into grand opera? Or nothing? A triviality compared to real life in a nation of 10 million, 80 percent of whom must scrape a living from the fragile soil, and cook their rice on fuelwood from the crippled forests?

Imagine that every museum in the world which has Greek statues sent delegates to meet each other on an island. Imagine that after three days they signed a paper saying that any statue which leaves Greece in future years must be paid in aid to help Athens cut smog emissions that corrode the Parthenon. And aid in both finance and training and knowledge for art and archaeology in Greece. And that Greek art in the future must be managed as a global whole. We stopped short of saying it would be inalienable Greek property but we have relinquished rights to buy and sell without regard to the whole.

And then imagine that the statues are not dead, but alive.

That they can fondle and groom each other, and mate and reproduce; that they can suffer, die, and be extinct forever.

In signing a plan for the future of little lemurs we have been talking about national sovereignty and a global heritage, and about the place of absolute knowledge and beauty in the world of money, food and firewood—and human pride.

After St. Catherine's the whole circus toured the country to see that zoos really do try to save endangered species. There was a hysterical moment in Los Angeles when the keeper pulled out a 200 pound python from its cage (a docile ex-pet called Lulu, though the keeper didn't admit that till afterwards). The Malagasy delegation, with long-established cultural terror of snakes, were challenged by the minister to hold it. Real, not simulated, giggles of horror. I wonder what it feels like if you are 16 feet long and eight people have hands under different parts of your belly and all eight are laughing in different rhythms? It would be dreadful if pythons were ticklish. The minister stood back taking pictures. Later he announced, 'I was the only one with the courage to say I was afraid.'

We were feted all along the way. First, sumptuously, by the Malagasy ambassador, Léon Rajaobelina (much more about him later). Also by Bill Reilly, head of WWF–US, who also invited all the Alton Jones millionaires, and by the lemurs of Duke University Primate Center, including a red ruff who lay on her back and invited Ratsirarson to massage her tummy. Out of the tour came a new creation by the zoos themselves: the Madagascar Fauna Group, which funds captive breeding in Madagascar, especially in the lovely zoo of Ivoloina, down on the coast near the port city of Toamasina.

Somewhere in all this Ratsirarson told us his history. Early in his career, the Water and Forests Department disciplined him for insubordination. He was sent to work in the remote reserves of Tsimanampetsotsa (scratchy spiny near-desert forest) and then Marojejy (near-vertical mountains sheathed in rainforest). Instead of pining, he fell in love with Malagasy nature—to the point of imprisoning all his father-in-law's village when they illegally cleared rainforest *tavy* patches two years running. Ratsirarson had long come across to foreigners as an implacable bureaucrat—now we finally found out that in the whole delegation he was the one most like us in truly delighting in life by the forest and in his passion for conservation.

Anecdotes and hospitality aside, there were two big impacts. No one at St. Catherine's had actually explained about cooperative management of a species.

Nov. 20, evening.

The red ruffed lemur studbook, compiled by Diane Brockman of the San Diego Zoo. As we dined in the cliff-top home of Margot Marsh, who has donated her millions to primate conservation, the minister picked up the red ruffed lemur studbook from Margot's sideboard. 'What's this?'

'Didn't they show you that this afternoon at the zoo?'

'No—the place they breed lemurs, that's all.'

So (admittedly with a bit of tugging from me) he and Diane sat on the cream tweed sofa in the nook with the enormous bronze African drum coffee table (Benin?) and Diane explained about individual tracing of animals and genetic management and the fact that all serious zoos are in it together.

And he studied it hard.

And asked sharply why this had not been presented in full at St. Catherine's. The Malagasy had no idea until dinner at Margot Marsh's what the zoo people meant by coordinated management of whole zoo populations. All that arguing about descendants, and no one had explained what a stud book is or why it matters.

The min. took the point. He now wants copies for all the Cabinet.

It was not until later that evening that he started to proposition Diane. He'd done the same thing two years before to Alison Richard, after officially opening her cherished university reserve, Bezà Mahafaly. My take is that if he met an attractive, forceful and highly intelligent woman he had to see how far he could dominate her. I wonder if he was surprised when each turned him down!

The other thing was the multimillion-dollar California condor breeding program—all that money and effort to save the last individuals of a species apparently dead set on going extinct.

May 21, 1987. The Condorminium at San Diego Zoo.

There are 14 California condors here, 13 in Los Angeles: the world population. The last free-flying condor, a male, was caught and caged on April 19, one month ago. There has never been breeding in captivity! One pair in Washington's National Zoo laid infertile eggs for years in the 1920s, but on dissection at death both proved to be female.

The California condor is big: black and white with the naked red vulture's head. We were allowed up to the shielded cage backs. We peered through little square holes in silence at a bird spreading its primaries on 10-foot wings. John Hartley whispered, 'This is primeval… how we are privileged…'

We shall not see them soaring on the thermals from the crags. The California condor in captivity is no more than a great hunched vulture. Andean condors circle up to 10,000 foot altitude, then launch themselves downwind, travelling at up to 100 miles an hour. They sleep sometimes in the evening 300 miles from the ledge they left that morning. The Chumash say their own souls must rise to the skies after death on the wings of the Andean condor. But the wild Californians dropped to a single breeding pair amid a lawsuit and emotional outcry to leave them in their wild dignity. Then the female died of lead poisoning. The judge threw out the lawsuit. The one last male has joined the few already in captivity.

Strangely, one apparently rational argument is all upside down. I had thought that the strongest reason to leave adults in the wild was for some to teach hand-reared birds the range and how to find food. Instead we must wipe out their present-day culture. The birds have been attracted to man for centuries. They eat aborted calves and lambs, wounded deer, the piles of deer guts left by hunters. Sooner or later they swallow a bullet—the lead from just one .33 is enough to kill them.

The goal, then, is to release the Californians in the remote Los Padres Mountains with carcasses set to lead them away from dependence on poisoned carrion. 'And must feeding in the Los Padres go on forever, unto all foreseeable generations?' I asked Mike Wallace, the curator of birds. He looked at me and answered, 'Yes.'[3]

The California condor story impressed the Malagasy deeply—for the incomprehensibly sentimental effort but chiefly for the millions of dollars spent, the apparently infinite largesse for nature by American zoos. Which brings us to a night in Saint Louis, in a mock-Tudor Inn, its walls hung with halberds and maces. In our endless late night confabs with the insomniac minister the one in Saint Louis was the climax.

3. The program continues. As of 2012 there were some 210 California condors flying free and 173 in breeding centers. http://cacondorconservation.org/programs; www.birdlife.org/datazone/speciesfactsheet.php?id=3821.

May 24, 1987. Tudor Inn, St. Louis, 11 p.m.
Voara circulated to say we were all bidden upstairs to a formal assemblage in the minister's room. The minister sat on his king-sized bed. John and Russ and the delegation and I picked chairs and sofas—Voara and Ratsirarson were left, so the minister waved them to also sit on the bed. They perched on the farthest corner, half behind him, as far as possible away.

He began by asking all our impressions, to summarize the results of the trip. Just informally, in the spirit of friendship—no, of a family which we have all become in this long voyage! He took off his jacket. Voara and Ratsirarson tried to edge even further away.

The minister then began to talk. He said there was indeed a new spirit abroad—a new and strict collaboration. Things would change from now on. Matters of the environment would be referred directly to him. There would be no more slip-ups. And there would be no more bartering lemurs for favors or any other reason. He might ask, for instance, how many lemurs someone accorded to Professor X in return for his doctorate. Ratsirarson sat in absolute rigor rather than flinch, and said nothing. Thereafter he tried to remain so immobile he wouldn't be seen.[4] (Barthélémy thinks he's in future trouble; I just thought the minister was demonstrating he could flick any one of us with the whip, at any time, as an example to the rest.) 'And Berthe,' he swung round. 'Do I understand you to say that letters must pass through the Ministry of Foreign Affairs—to you, to me, to Mr. X or Mr. Y? Or only official correspondence? Because I will not be cut off from information. The president of the Republic would not have sent a minister and a

4. His doctorate came from the well-known and respected lemur genetics group at the University of Strasbourg.

secretary general and all of you as well if he did not attach extreme importance to these questions. Speak!'

Mme Berthe was tough and forthright. She said she only meant official letters, not all the letters, of course. Having attacked Mme Berthe, the minister suddenly agreed.

He swung round to Barthélémy. 'And you—what do you say? I welcome you to our group, in which you have played a full part.'

'Indeed,' Barthélémy said, 'It is very important to have had this chance to talk.' He explained to John and Russ, 'We are almost neighbors in Antananarivo, but I had to come to California to meet the minister for discussion face to face... And now that the channels with WWF have opened, I would like to come to you directly when there seems to be a need, not stop in the outside office of someone who has no power of decision.'

The minister let him talk a little too long, until it sounded as if Barthélémy were expostulating in a vacuum. I don't think he agreed or disagreed when Barthélémy tailed to a stop.

All of this was preamble, not really to decide anything but to show that he, the minister, was in charge of any powers of decision. The cowering members of the delegation must obey him. To do what?

At last he came to the point. The minister slapped down onto the bed Russ Mittermeier's Environmental Action Plan, with the map on the front and ring-tailed lemur on the back cover, several hundred loose pages in a big black clip. (Russ had told me how he'd been played cat-and-mouse with yesterday, when he tried to present it to the minister, who pretended total lack of interest because he could see how much it meant to Russ. Russ had no idea whether all his labor would be instantly buried.)

'This,' said the minister, is a Plan of Environmental Action for Madagascar. It is a five-year plan covering all proposed projects. How shall we go about having it accepted?'

(One sensed, rather than heard Russ gasp—he wouldn't gasp in this kind of poker game.) 'It's only a draft, a preliminary proposal' he put in quickly. What I hope for is your comments and ideas at this early stage.'

'Exactly—now, can we have it sent to the inter-ministerial committee: the persons here present, who have undergone the intimacy of St. Catherine's?'

People can smell fear—not just dogs. No one wanted a committee of themselves responsible for an action plan, let alone action.

'Very well—as I see it, there are three choices. First, it could go to me as head of this committee, to disseminate to you for comments. Second, WWF–US could finish it and send it as an official proposal written by *vazaha*, foreigners. Third, this draft could go to Affaires Étrangères to send to everyone in the government who thinks he has an interest in the subject to ask for all their input. What do you choose?

The second choice was an obvious dummy: if sent cold, the plan would be slapped back cold to Washington, and the minister had said he wanted it accepted. That gave everyone a brief chance to agree. But there the real choice lay—did they include their own ministers and ministries, or did they shift allegiance to this new committee and its dangerous chief?

They tried to find ways out, but he leaned forward and said, 'After all, it is my responsibility. I am minister of Water and Forests, as well as all our nation's 10 million cattle and the fish of our lakes and seas. The reserves and the forests and lemurs are mine, in the end. Don't worry, I am like that (two fingers together) with Rabesa, your minister of scientific research, like twin brothers, and like that (three fingers) with Rakoto Ignace of higher education. The document will go as a draft to this committee; we will modify it, and then it can be presented as a work of collaboration between Malagasy and *vazaha* to all the others concerned. With the president's help, I think we should pass that final stage in about five minutes, a few months from now.'

All this was in French, sealing the four *vazaha* to the others as witnesses and accomplices. Then he sailed off into Malagasy with its rises and falls and checked pauses and swooping phrases like a hawk that stoops on the listener's mind. It wasn't long, but it was final.

Later the Anglos reeled off to the pseudo-Tudor pub bar and rehashed the whole thing, as the Malagasy doubtless did in their own corner, while the minister probably forced the secretary general into yet more Scotch. Russ wanted to know if you could hear his heart thudding across the room when the minister pulled out his Action Plan, and I wanted to know if any of them had ever met another person quite like that.

EIGHT

The Bank corrals the donors

Russ Mittermeier's draft Environmental Action Plan was mainly about forests and lemurs. In 1987 Barber Conable, president of the World Bank, had announced that the Bank would undertake country-wide national environmental assessments. Madagascar was the first country to take up the offer. Léon Rajaobelina, the Malagasy ambassador to the USA, applauded the move, but stressed, 'We are tired of being studied from the outside. What we need is a truly Malagasy plan which leads to concrete actions on the ground.'[1]

Léon Rajaobelina was not an ordinary ambassador. He had been governor of Madagascar's Central Bank after working in the IMF. Madagascar posted one of its brightest and most influential economists in the key role of negotiating with the Washington institutions which held the purse strings of the Malagasy economy through the overwhelming burden of debt.[2]

A leading World Bank environmental economist, François Falloux, and his colleagues spent the next two years drawing up a much more ambitious document. It tackled questions

1. F. Falloux and L.M. Talbot, *Crisis and Opportunity: Environment and Development in Africa* (1993). p. 31.

2. For Léon Rajaobelina's account of how Madagascar sank into debt, see A. Jolly, *Lords and Lemurs* (2004).

from fuelwood to soil conservation, environmental policy, an environmental information system, land tenure, urban and rural environment, wetlands, biodiversity, ecotourism, environmental research and development of human resources. In 1988, Falloux led a large mission to Madagascar to consult widely with people there.

By 1989 the multifaceted Environmental Action Plan had been drafted. It was usually called the EAP, but people tried to remember to say NEAP: National Environmental Action Plan. The aid agencies funded a NEAP support cell in Madagascar of some 150 Malagasy experts to give national substance to the EAP. Falloux writes: 'The government had the wisdom not to entrust the NEAP preparations directly to the existing government structure ... even though the CAPAE [the support cell] was sponsored by the Directorate of Planning.' The ambivalence in this wording for a donor-funded agency outside existing government structures reflects the disparity between donor and government influence that dogs the whole history of the NEAP.[3]

Now the Plan would need multi-donor backing.

I continued to be amazed to be invited to such meetings. Berndt von Droste of Unesco, mostly concerned with the education facet, seemed to think I had something to contribute.

January 27, 1989, Washington DC.

On the fourth floor of the main World Bank building, past the flags and the front door guard, up the teak-colored elevator, past the great square central hall hung with photo-realist murals of parrots and women and tropical flowers, into conference room E with coffee urns in the anteroom and a glimpse of curved teak table beyond.

3. Falloux and Talbot, *Crisis and Opportunity*, p. 30. The support cell was officially the CAPAE, the Cellule d'Appui au Plan d'Action Environnemental.

We are milling around drinking coffee and greeting: gray men in gray suits. There is only one black delegate, a silent African from UNDP. And one woman, a tall, elegant German with militarily short hair, a bright crimson dress, black sweater, belly-length pearl necklace and crimson lipstick—the kind of red and black that is more aggressively executive than gray (she didn't actually say anything though). In fact, the softest things in the room were Chuck Lankester's curly beard, and François Falloux's glorious spade beard and parted mustachios, like a De Brazza's monkey.

Russ Mittermeier is older. The gray streaks are taking the color from his hair, which is short, not the flowing brown mane of two years ago. He has a mustache: a straight bristly line, though he fingers it and says he plans to outdo the Kaiser Wilhelm curlicues of emperor tamarins. And he is much thinner, with sharp hollows under the cheekbones, instead of beefy musculature.

He does a double-take when he sees me, too, then recovers by saying I'd put my hair up. In fact, I had bought a new suit: beige and black business tweed with a discreet red stripe, and an ecru silk blouse. Mother's pearls make me completely effaced, with hair more pearl now than brown. A perfect disguise for attending the World Bank, I thought.

We filed in and sat down at the curved table, twenty to a side, with Hans Wyss of the World Bank chairing, Madagascar's Ambassador Rajaobelina at his right hand, Russ at his left. I took one of the observers' seats, behind Berndt von Droste of Unesco, who had invited me. Officials had place cards.

Rajaobelina opened the session by saying he must pay tribute to two people who had done more than anyone else to promote conservation in Madagascar, over the past twenty years: Alison Jolly and Russ Mittermeier. I found myself grinning like an idiot, looking round the room at each person like a ballerina bowing with her bouquet. Berndt turned round in his chair and clapped silently.

Then the ambassador opened formally. This extraordinary initiative honors Madagascar, and reflects the country's political will to conserve its environment. The president of Madagascar himself takes an interest in the outcome of today's meeting. Above all, though, the program we are about to launch must be under Malagasy control. It must be Malagasy in reality, not just in name.

L. Sayers, USAID. Madagascar is one of the United States' highest priorities. (Imagine anyone saying that five years ago!) Future funding for the environment and the new *Centre Nationale de la protection de L'environnement* is now assured. The main task now is to get it right, and to avoid false starts and failures, because the eyes of the world, and of the population of Madagascar, are all on this grand initiative.

Bernard Pasquier, the World Bank Country Officer. The Bank now has a $1 billion portfolio in Madagascar, and adds $100–$150 million every year. (Translation: $1 billion of Mad.'s $2.6 billion debt is owed to the Bank, plus interest.) No other sectors will succeed without the environment, e.g. dams, and irrigation projects in a place where 80 percent of the population is rural and most of them eat irrigated rice. Mad. loses $250–$300 million per year from erosion = 10 percent of GDP. It also pays more than 50 percent of export earnings for debt service. Needs massive debt-for-nature swaps. The Bank's environmental commitment is firm, and projected for twenty years.

François Falloux, World Bank architect of the Environmental Action Plan. Madagascar now retains 16 percent of original forest cover. It will be zero in 2030 or 2040. The first tranche of funding will be $60 million for 5 years. But it will be spent with two-thirds for 'development,' one-third for biodiversity. This is a process not a product. The goal is sustainable development for a whole society.

Is it just with hindsight that I wonder about what constitutes 'development' in Falloux's list? Land surveying and entitlement, marine monitoring, reforestation in the watersheds of the electricity-producing dams and the rice bowl of Lake Alaotra? In later years the Bank would stress eliminating poverty and even providing rural livelihoods, but in 1987 it was still top-down thinking—and of course loans, not grants. It was only Unesco which said anything about education.

Berndt von Droste, Unesco. We also need primary school teachers, learning packages, women's education, science and technology, especially biotechnology, photovoltaics. Marine science: training scientists and institution-building. Look back thirty years to the pioneering foundation of the Charles Darwin Institute on the Galapagos. I put in my hobby horse of working through the Malagasy Ministry of Education rather than creating parallel structures under the pressure of the Bank.

Chuck Lankester, UNDP. UNDP has put $¼ million (out of $¾ million) into a feasibility study for how to raise funds on the necessary scale, including massively expanding debt-for-nature swaps rather than just adding new debts. Will we really look back in thirty years to today as we do now to the founding of the Charles Darwin Institute? But we must be sure to learn the lessons of the past. UNDP's Operation Savoka has dribbled along for years as a pilot project until UNDP gave it up. Did it really wean farmers from slash and burn? Did it really offer alternatives? Why didn't it spread, if so? (I could tell him a lot more things that didn't work. If *only* I knew enough to foresee which bits of today's grand hopes won't. Or maybe I don't want to know.)

Ken Piddington, World Bank Environment Section. He sees Mad.'s political will, and no generic difference between First and Third Worlds in the need to enlist support of populace round parks. (He's a former New Zealand national parks administrator.) No word of his fears, conveyed elsewhere, that there is far too little sociological understanding in the EAP, echoing Chuck on the Savoka project's lessons.

Russ strides round down the far end of the hall to the slide projector which has been waiting on its industrial stilt table between the two teak wings of conferees. Lights down, slides up—a massive baobab (*Adansonia za*, I think, from our side of the island), the standard ring-tail which has been on so many covers, his adorable mouse lemur I always start lectures with, my own slide of fossil giant lemur skulls. And statistics: highest proportion of endemics of anywhere except New Caledonia, 293 of 297 reptiles shown over a sunlit radiated tortoise, half the bird species, with that *Coua cristata* and its magenta and turquoise eyepatch. And so on. It's funny to know by heart the view of the side of Nosy Mangabe's rainforest, of the de Heaulme Aepyornis egg, of that mother and son golden bamboo lemur at Ranomafana, where other people just see a succession of unpredictable marvels. Russ ends that Madagascar right now, for richness and vulnerability, really is the first priority of all.[4]

After lunch, sums of financial aid promised by assembled agencies. Then late in the afternoon an aide passed a note. Ambassador Rajaobelina rose from his chair and positively

---

4. The game-changing article by Norman Myers, Russ and others on biodiversity hotspots must have been already in press. They argued that 1.8 percent of the world's land surface holds most of the world's biodiversity, and that the top twenty-five of them hold 44 percent of flowering plant species and 35 percent of vertebrate species. Of these, Madagascar comes out hottest of all if one adds in the percentage of habitat loss. N. Myers, R.A. Mittermeier et al., 'Biodiversity hotspots for conservation priorities' (2000).

scuttled to the anteroom, amazing in a man whose every move, as well as his mind, conveys penetrating sophistication. He then announced to us that the president of Madagascar himself had telephoned (at midnight Madagascar time) to ask the course of the meeting. Rajaobelina closed the session with graceful words from the chair, and then swept out, his aides behind him, with the speed and poise of a president himself—or a man who has guided his country's next step to the future.

NINE

Dishing out the dough

In July 1888 the World Bank issued a draft Environmental Action Plan for Madagascar, with the help of a donor-funded support cell based in Antananarivo which called on some 150 Malagasy experts and technicians. The chief architect of the EAP was François Falloux, a leader of the Bank's environmental section. Actually it was not only the Bank, but USAID, Swiss Coopération, Unesco, UNDP and WWF—an NGO among the international and national aid donors. A major multi-donor mission in March 1988 had pulled all this together for the July draft.

The EAP was not just about biodiversity. It foresaw five priority areas: biodiversity, community environments, soil erosion, land tenure and education, including training and public information. Not just the multi-donor involvement, but the actual scope of the EAP was pioneering for the donors: an attempt to deal with the whole interface of people and environment in a rural-based country.[1]

Education was already under way at WWF. The indefatigable Barthélémy Vaohita wrote a whole series of primary-school readers titled *Ny Voara*, 'Nature,' which were

1. World Bank, *Madagascar: Plan d'Action Environnemental*, Volume 1: *Document de synthèse générale et propositions d'orientations* (1988).

distributed to the provinces in 1987. Some teachers used them, but the Ministry of Education still couldn't see the point of readers and storybooks about the environment. It considered learning to read and cipher quite challenging enough without muddling children by adding environmental messages to the curriculum. At least one teacher was found (in 2010!) using the still-wrapped copies of *Ny Voara* as a stool because she had no chair.[2]

The most controversial part of the EAP spelled out plans for dealing with land titles. All rural reform efforts in Madagascar come up against intricate and entrenched confusion in traditional land rights. The system is 'involuted,' Clifford Geertz's classic term for the layers on layers of ownership he found among the paddy-rice farmers of Java: kinship obligations, sharecropper obligations, obligations between ex-slaves and ex-slave holders.[3] Often people do not want clarification. A legal title could lead to someone else having the right to sell the land, even land that holds one's family tombs! François Falloux, the report's chief author, was an expert in land tenure. He hoped the EAP would be a chance to set Madagascar on some clear-minded course, ensuring more formal ownership by those who actually work the land.[4]

Although estimation had to be approximate, the 1988 draft EAP envisaged some $300 to $400 million over 15–20 years for all the proposed components. This might be split as some 30–40 percent for physical infrastructure, 20 percent for conservation and management of biodiversity, 10–20 percent for human resources, and all of 25 percent for surveying and titling of land. This left 5 percent for overall administration. However, set against this was the calculation that Madagascar

2. F. Falloux and L.M. Talbot (1993). *Crisis and Opportunity: Environment and Development in Africa*; Prof. Hanta Rasamimanana, pers. comm.
3. C. Geertz, *Agricultural Involution: The Process of Ecological Change in Indonesia* (1963).
4. Falloux and Talbot, *Crisis and Opportunity.*

lost annually between $150 and $300 million to environmental degradation, or 5–15 percent of its GDP. This was enough to convince Prime Minister Victor Ramahatra and director of planning, Philippe Rajaobelina, that the investment was worth it. They even, but more slowly, convinced the skeptical president.[5]

The last chapter's meeting in the World Bank in Washington was a day of bidding up funds. Now came the nitty-gritty: what the Bank calls an appraisal mission. You might think that with the enormous effort already put into the draft EAP, and a previous multi-donor mission with many of the same people in 1988, there was no real need to do it all again, but it is the appraisal mission which actually sets out how funds should be allocated within the country. The members of the appraisal mission comprised about twenty foreigners and two outnumbered Malagasy. I was sent by Unesco to draw up the sector of environmental education.

Our main interlocutor within the Malagasy government was Philippe Rajaobelina, director general of the Planning Office and brother of the indomitable ambassador. Even higher, prime minister Victor Ramahatra strongly supported the EAP. Notably absent was the Napoleon of St. Catherine's, the minister of Water and Forests. He had transferred to the Food and Agricultural Organization of the United Nations in Rome. I have absolutely no basis for this speculation, but I wondered whether the president of Madagascar thought Rome just about far enough away to send a minister so nakedly ambitious. If so, the whale ate Ahab.

Meanwhile, the donors were clear that the future reserve system should be under a separate agency, not Water and Forests. This might defuse the conflict between forests saved for biodiversity and forests used for profit, with all the

5. Ibid., p 43.

temptation for people to line their pockets from one of the few 'free' resources. This would also allow the new parastatal agencies to pay higher salaries than regular government, poaching the best people from Water and Forests.

The NEAP created a National Office of the Environment (ONE), with a sweeping mandate to coordinate environmental policy. The National Parks and Reserves went under ANGAP, the *Association Nationale pour la Gestion des Aires Protegées*, to the fury of Water and Forests. Finally, ANAE became a nationally based funding organization for smaller projects carried out by smaller organizations.

The end result was to be a new National Environmental Action Plan, the NEAP: offspring of Russ's draft EAP of the great Tudor bedroom meeting two years before and François Falloux's meticulous planning. Jean-Roger Mercier, looking back on the preparation of the NEAP, writes 'Our enthusiasm was probably in direct proportion to our political ignorance of the depth of internal struggles within Malagasy society, and our equally naive perception of the ease to transform additional financial support, which seemed to be ready to flow towards Madagascar, into improved environmental protection.'[6]

July 10, 1989. Antananarivo. Letter to Richard Jolly.

Let me try to write you what really happened on the World Bank appraisal mission. Then maybe I'll find out myself. The first two weeks were research, meeting people in our sector, and uniting in the evening to go over the day's adventures and have our in-group confabulation about how the mission is going. Picture us in the presidential suite of the Hilton, sitting (all 20+ of us) round a U-shaped table

6. J.-R. Mercier, 'The preparation of the National Environmental Action Plan (NEAP): was it a false start?' (2006).

with a white cover and a red frilly skirt. Windows on the long side look out at the Queen's Palace on its hill, gold with sunset.–

François Falloux is at the head of the table with his spade beard and neatly parted mustache, and a shirt with wide blue and white stripes. On his right is his right-hand man, Bruno Ribon, an abnormally thin, saturnine Frenchman who is head of the Bank's biodiversity team. Ribon explains the Bank's policies of cost calculations to the rest of us with the aid of felt tip pens and a flip chart. He could play Cardinal Mazarin.

No one but Bruno will sit at the exposed position beside François, so there's a space to François's left. Then, decorously round the corner of the table is Viviane Ralimanga, head of the support cell to the EAP. Viviane is about 30, with quick grin and close-cropped hair. She graduated from the University of Madagascar as an agronomist and did a Master's in agronomy in Louisiana. She's been working on the EAP for a year and a half as head of its secretariat. She turns out to have a mind of her own.

Opposite her, on Bruno's right, sits Daniel Crémont. White hair, erect military carriage, the slight garrulity of the distinguished retiree. He does satellite mapping and surveys. François confided with some awe that Crémont negotiated a $250 million loan to Brazil for satellite surveys, so Madagascar is peanuts.

The rest of us are higgledy-piggledy down the table: Jean-Roger Mercier, energy, information bank, cost-table coordination. Oliver Langrand, birds, and Martin Nicoll, mammals, here for the last three years surveying reserves for WWF. Roland Albignac, thirty years in and out of here, now working for Unesco in the new aye-aye biosphere reserve at Mananara. Arild Oyen of Norwegian Aid, twenty years here; Klocke of German Aid, two days here. Frank Convery, Irish economist, also first time. Jan Post, World

Bank and marine ecologist. Hans Hürni and Kuno Schlaefli, soil conservation from Swiss Cooperation via years in Madagascar and Ethiopia. David Gibson, forester, USAID from Nairobi, his head cocked aggressively upward. Lee Hannah, USAID Washington. We were sometimes joined by Donna Stauffer, USAID Madagascar. (Russ Mittermeier should have been here, but he isn't. He has just spectacularly left WWF, absconding with all his henchmen to something called Conservation International. No one knows what will happen to WWF's Madagascar grants.) Kathy Richman, executive secretary, will coordinate and write the final document. She's a consultant hired in by the Bank just for the EAP. A cheerful, plump girl from Cambridge, Mass. She needs to be cheerful.

François says it's the largest mission he has ever worked with. At the end, he said it was the hardest one.

We didn't even finish. There'll have to be another mission in January.

Did you notice there was only one Malagasy? Oh, and our sociologist, suave Prof. Ramonjisoa from the University, but he didn't say much.

Saturday, July 1, 1989.

The weekend retreat in Perinet was a drizzle-out. We went down to Fanalamanga on Saturday (7 a.m. start on the sick-making Tamatave Road—deep potholes again since the Chinese stopped fixing it last year.) Fanalamanga is the doomed World Bank scheme to plant 200 km^2 of pines and make Madagascar self-sufficient in paper pulp. They planted them all right, and kept the fires out by a mobile helicopter firefighting patrol. Only the soil wasn't right, and the pines wouldn't grow on schedule, and now they're scraping around for a controlled way to let local charcoal burners convert it to kitchen charcoal.

Lac Alaotra's irrigation was another World Bank scheme. Do you remember when the Bank agricultural chief Monty Yudelman took up his post he said he wanted to start by seeing a country where not one World Bank project had succeeded? So they sent him to Madagascar.

Saturday noon. The agricultural saint.

The Bank mission escaped from Fanalamanga's huge and ill-thought pine plantation. It's Saturday, the sun has come out, and I trail the gang up a half-kilometer-long path to arrive in mid-explanation. A short man in bare feet and a tattered old sweatshirt with mock-computer lettering is lecturing to our gang under the eaves of a wooden house. I have no idea who he is, but he is using French with verbal flourishes and scientific precision—I keep wondering why he can't afford shoes.

He explained about the trials of agroforestry with *Grevillea, Albizia*[7] and other trees to speed up and settle the *tavy* rotation. And how hard the peasants work in a place where 60 percent of the land is at more than a 40 percent slope, and it rains 260 days a year ('Plateau people say *tavy* farmers are lazy, but if you think so, just you try working in these conditions!') and how much he'd learned about community and clan traditional land rights where outside people see unclaimed forest.

This is the man to explain what 'slash and burn' *tavy* cultivation really means: to peasants, and to conservationists, and to people who think land tenure will be solved with the Bank's great big satellite mapping machine and its attendant technicians all sitting in Tana. He hopped up and down the hillside showing us tree plantations and waving

7. *Grevillea robusta* and *Albizia procera* are widely used agroforestry species.

his hands, his French leaving me more and more weak with envy.

'Who is that!' I hissed at Jean-Roger Mercier, most knowledgeable of the Bank team, as we trooped off to the bus.

'That is Jean-Louis Rakotomanana. He could work in any country in the world, and he stays here,' was all Jean-Robert would say. 'I think he's a saint.'

Jean-Louis turned up in Tana next day, with shoes and a somewhat more recent sweater, and came to lunch at the Hilton. The last soil science conference he went to, in Bangkok, had suggested very useful parallels to the Malagasy *tavy* cultivation. The African conference in Rwanda had less to offer, but as he was rapporteur he'd put in some of the Malagasy cases. And I lived in New York? He'd love to see New York again, but hadn't for twenty years—not since his post-doctoral studies in California...

Yes, he'd be happy to join the committee on how to communicate with the farmers, if Ramahatra asked him, but Dr. Roland Ramahatra was a great politician, and he, Jean-Louis, was only a small, small person—a research scientist. But he'd be glad to help.

Saturday afternoon.

The great internal self-education of the weekend retreat. Flip charts set up in an overly large, bare room with bare wooden tables. The flip charts leered at us, with green felt tip and red underlining. Furthermore, it was Perinet winter weather—interminable fine grey cold drizzle outside the cracked and broken and boarded-over window panes, and colder indoors than even out in the rain. Today and tomorrow everyone presents their sector. I am too cold to concentrate.

I cheated, and brought a whole bag of children's school books and adult education books and from many countries, a packet of Pied Crow environmental comics from Kenya, and everything under way in Madagascar. I spread them out on the conference table and caught several delegates reading kids' books when they should have been listening.

There doesn't seem to be much argument about the education plan. The chief surprise for me is how much is already going on—NGOs, WWF, major Bank education programs, Swiss overhauling the radio broadcasting, a Unicef program, German Aid basic reading manuals. Nobody in the education game tells you about the others. I was still on a steep learning curve when I left. What I didn't do was put in nearly enough about bottom-up communication. The other aid donors' counterblast got me for that, saying there should be as much listening to villagers as talking down to them.

I am not sure whether I was just too cold, but very little of the other sectors made any sense to me.

Monday, July 3, 1989.

After the weekend of internal presentations in Perinet, Hans Hürni gave François the black spot. I said we were seated higgledy-piggledy. That was in the first two weeks, before our weekend in the rainforest telling each other what we thought all our components were doing. Afterward the aid donors sat in a grim clump at the far end from François. … When François called on Hans Hürni, Hans just rose with a paper from the donors in his hand, walked silently the length of the table, put it down in front of François and walked silently back.

Do you remember when the mutinous pirate gang in *Treasure Island* presented Long John Silver with the black spot? Of course a memo querying the leader's leadership would have been lethal for Long John Silver, but was still plenty embarrassing for François.

Only approximately 20 percent of the nascent Environmental Action Plan is for biodiversity. That is, $21 million over five years, of which $13 million for reserves and protected areas proper, $9 million for eco-development around the reserves. All of this to be under the new parastatal agency, the National Office of the Environment, not the wholly corrupted Eaux et Forêts. Martin Nicoll and Oliver Langrand have been working for three years toward this. They have listed reserves in order of priority, and needs, and are ready to roll. USAID has backed them up, and has its own ideas of micro-irrigation schemes and other 'eco-development.' Of course no one has solved the social problems. (I don't even know if any biosphere reserves work anywhere, outside of England's national parks or the Cévennes park in France, which are biosphere reserves with people and landscape blended pretty well together.) But they have a plan.

The Aid Donors' memo, Hans Hürni's 'black spot,' rightly stressed that biodiversity is the one coherently worked-out section, the original heart of the project. Their memo suggested cutting everything else down to pilot programs and concentrate on the biodiversity aspect that started the whole EAP. USAID announced it has dibs on the biodiversity section.

François read it, and then read it out. It would mean dropping all the other sectors: land surveying (of which more later), erosion control for the Lake Alaotra rice fields, and a gaggle of smaller pieces that were on the table, including my environmental education section.

By the next day the Malagasy counter-attacked. Viviane Ralimanga herself wrote an impassioned letter saying if we thought we could just emphasize fauna and flora, we were sadly misjudging the temper of the Malagasy, as well as their needs. Philippe Rajaobelina, directeur générale adjoint of the Plan, wrote a letter saying that even within the biodiversity section it was unacceptable to have more money allotted to the reserves than to peripheral development, because 'There are more important primates in Madagascar than lemurs.'

No one made the point that the Bank estimates Madagascar will get $700 million in aid and loans next year. Even supposing $300 million of that is fictitious, merely paying the debt interest, it is still churlish to begrudge $4 million next year to nature protection, especially since Malagasy nature is the one good thing which the world at large knows about the country.

But passions are high.

The next biggest section is some $25 million for teledetection, land surveying and land title tribunals. The first step is to buy some $10 million worth of equipment. The next step is to sit on two warring agencies in the government, because only one of them can have that equipment. After some institutional surgery, one then trains 500 surveyors to run the machines and also a few high-level administrators to oversee it all. Out of it comes, with luck, a set of base maps, a set of roving land courts, and enough land titles granted five years hence to equal what would be granted under the present system in eight centuries. (There is currently a century's backlog in survey and title requests.)

Poppycock, say the other aid donors. For all Daniel Crémont's august and snowy-haired demeanor, he and François are offering technicians' pipe dreams. It will all bog down in Malagasy deliberate inefficiency. No one wants the land titles clarified except people who don't currently have land titles, and who do you imagine will listen to them?

'You don't understand,' replies François. 'There is accelerating land-grabbing now in the west, with the fat cats fattening. On the plateau, much belongs to absentee landlords, farmed by sharecroppers on short leases who have no incentive to build dykes, save soil, plant trees. Only rapid registration of titles can save traditional rights to land, and avoid it all going to whoever can pay for a survey.'

'Fine, register it,' says Hans Hürni. 'But you'll need land courts and adjudication, not fancy maps. Make your judgments on the spot, and put a great big stone at the field corners. Your $10 million machine would do more good holding down an agreed field corner than it will in Tananarive.'

'You're talking about land reform!' That wasn't said in the mission. It was said by Dr. Roland Ramahatra, president of Tananarive province. He said it to me while we were driving back from a glorious day among plateau scenery, visiting the villages of the Swiss agroforestry scheme, 20 km south of Tananarive. The Swiss had been saying (everyone says) land tenure is their single greatest obstacle. The villagers we met were enthusiastic about planting cherry trees and pine trees—if they protect the seedlings and saplings for seven years, they gain title to the land. However, jealous neighbors have been known to burn down the planted pine tree stands in year six. Only Dr. Ramahatra is blunt enough to name what would be really necessary to deal with land ownership here as serious land reform.

And I keep wondering if the Bank's accelerated scheme won't just help the people who know what's happening and when there will be a land tribunal to handle it, the fat cats will grow into tigers.

Yet another bone of contention: soil conservation and Lake Alaotra. Hans Hürni sees soil conservation as micro-projects to be run by NGOs that the Bank or EAP would finance. François, though, promised we'd do something about the watersheds of Lac Alaotra (Madagascar's rice bowl). Hans's professional opinion is that Lac Alaotra, at least the south-eastern part, is a write-off. Huge *lavaka* silt up rice paddies as fast as they can be cleared. Lavaka are erosion gullies in red laterite soil: red clay stained with rust (iron oxide). The soil's top crust bakes hard in the sun but the soil beneath is soft, so when rain comes it gouges out a huge gully in the side of the hill as pieces of crust fall into the crevasse below. There is dispute amongst geologists how much erosion dates from earlier periods of high rainfall and how much is due to recent tree clearing and grass burning by people. Some lavaka are clearly stabilizing with vegetation growing in the reddish gash, but it is difficult to stabilize them this way on purpose. People plant trees at the top, but the lavaka are growing on a scale no mere root system could stop. Anyway, upland pastoralists who have no stake in the rice fields burn down trees as fast as they're planted. French Cooperation is now interested in heavy bulldozing and concrete retaining walls for the lavaka: Hans says if they want to take on major construction, OK, but let's keep sensibly out of there.

At this point Jan Post burst out, 'If my country were sensible about the effort it takes to make land, we wouldn't have a country! The value of Lac Alaotra isn't a dollars and cents cost accounting, it's the country's rice production! Do you think the Netherlands could afford to listen to a cost accounting when we built the dykes?'

To summarize: Malagasy versus all foreign donors on how much should be allocated for biodiversity versus other big projects thrown in with the EAP. Aid donors versus the Bank on whether there is any point in heavy-duty satellite mapping and land registry at high cost and little communication with peasants. Aid donors versus the

Bank and each other on whether to take on big projects of soil conservation, especially the rice bowl of Lake Alaotra, within the EAP. We were supposed to have sorted all this out on our weekend 'retreat' in the rainforest. And everyone is paying lip service to bottom-up communication while proposing zilch about it.

Monday, July 2, Antananarivo.

I know who put them up to bottom-up communication. I realize the fundamental difference between Dr. Roland Ramahatra, president of Antananarivo province, and all the rest: he's an idealistic politician. He's also an intellectual and an aristocrat and a remarkable man, but the ultimate difference is he's an idealistic politician. He actually wants to go to the people and stir them up and ask them questions and give them things—maybe even land. He backed the EAP support cell—Viviane's gang—to hold a seminar for notables from each *fivondrona* (old sous-prefecture) of his province to come and voice their views on the environment. (Of course they all know it's deteriorating.) In fact he has, or had, high hopes for the EAP as at last a program that would reach people, not just the administration.

So did his brother the prime minister, who said so in so many words to François.

One thing which queered this approach is precisely our earnest Malagasy collaborators—the 150 civil servants assembled in working groups to give a genuine Malagasy viewpoint. Of course they're all administrators, and 99 percent plateau people. Most of them have never been in a forest and privately recoil at the thought of living near one, let alone listening to the kind of people who live next to forests.

Arrived back at the Hotel Colbert Sunday night to find they'd put a sleek, space-age, streamlined telephone in my room, with instructions to dial 9 for operator, 52 for breakfast and 51 for reception.

Of course it didn't work. No dial tone.

So, Richard, I stood downstairs for an hour on Sunday night phoning you as well as leaving instructions for café complét avec extra milk in my room at 7.30 Monday.

At 7.15 a.m. they hammered on the door with a note to call Dr. Roland Ramahatra IMMEDIATELY. I flung my Himalayan shawl as concealingly as possible round my nightgown (your Sudanese sheik's robe) and jammed on sandals, and fled downstairs to phone the genial provincial president.

He was upset.

He'd heard there was far too little bottom-up communication in the Plan. Not just my part, but all of it. Furthermore he'd heard we might ditch Lac Alaotra. He hoped I'd use my influence (!) to make people see that Lac Alaotra is very important to the Malagasy. He went on and on, constantly returning to my lemur poster (*Arovy izahay*—protect us!) and the mournful eyes of all the lemurs above his desk pleading with us to get the EAP to work!

And I was alternately annoyed and giggling, hiding as best I could behind the door of the Colbert front hall telephone cabin, in my nightgown and dinner shawl.

Friday, July 16, Antananarivo. So what has happened? François has bluffed it out, much as Long John did with the black spot. ('This 'ere paper's been tore out of a Bible! What unlucky sailor's gone and spiled a Bible to give me the black spot! I'm afeared fur you, whomever done it...') We staggered on through the week to the final meeting on Friday with a roomful of Malagasy chaired by Philippe Rajaobelina, Adj. Dir. Gen. of the PLAN.

We each presented our sections of the EAP. Rajaobelina still said there was too much for biodiversity. A great silence greeted the $25 million land survey plan, except for a kind of dying squawk from the Direction de Patrimoine (the land heritage unit), which didn't get the $10 million machines—but someone else still does. Lac Alaotra sounds as if it's still in, because it will get lots of mini-projects for soil conservation—but I don't know how long that will fool the Malagasy as a solution to ravaging lavaka.

Having carefully read the World Bank's massive proposed education loan, and the gradually forming EAP, there are some common elements. (1) Pots of money. (2) A real, desperate need. In the EAP's case, the soil really is washing away, and the future's fuelwood burning up. There's no sign Madagascar will cope if someone else doesn't do it for them. The same with education—schools without books and teachers without training and closed teacher's colleges are a scandal. Maybe this all happened because of structural adjustment, or whatever, and the Bank and the IMF knocked them flat before helping them up again, but only the Bank seems remotely big enough to help now (CRESED, the education program, is $30 million). (3) The strategy includes administrative reorganization near the top. One of CRESED's conditions is the creation of a Unité Pédagogique, or curriculum reform unit, with sweeping powers. The EAP is proposing a 15–20 person National Office of the Environment. The EAP excises nature reserves from Water and Forests and does major surgery on the Direction du Patrimoine. In other words, a fast way to start things working according to the Bank's ideas is to create a new executive body free of previous ministerial deadwood or control. (4) Existing governmental interests don't like it.

I turned the tables momentarily on the genial Doctor Ramahatra by phoning him the day before (fully dressed) and asking him to chair a committee to figure out how to listen to peasants, because I couldn't figure it out. People liked my presentation, if only because I said it would all have to work through established ministries. They even said, 'Can't Education start early without waiting while we argue over all the rest!'

After Friday morning's stand-and-deliver, the 'closing pre-lunch cocktail,' and the closing mission meeting, François Falloux actually hugged me (not for my whole-hearted support—I think I'm just the most huggable mission member) and confessed this has been the worst World Bank mission he's ever been through.

So I said, 'François, I've got just one question. Isn't what the Bank is doing the new colonialism?'

'Of course. Exactly.'

So I tried another: 'In how many countries has the Bank taken over virtually every ministry, like Madagascar?'

'Not many. Ghana, perhaps. It's more partial most places.'

'How do you justify it?'

'This time we have to get it right.'

Environmental education never did get funded in the first EAP. The African Development Bank was interested for a while, but that flopped. The Malagasy government didn't want another loan with no immediate return, in any case. I followed up other aid donors the next year, but they thought the possible results too far off and too iffy or were concentrating on other education programs. Instead the WWF stepped in with a widely popular teenage comic called *Vintsy* (Kingfisher). Other NGOs such as the Madagascar Fauna Group and the Wildlife Conservation Society have promoted environmental education, but only in the local regions where they work. That is why my colleague Hanta Rasamimanana and artist Deborah Ross and I made the little Ako series of lemur books (see Chapter 4) as a further pebble of empathy to toss into the pool of indifference. Someday all Malagasy primary schoolchildren will learn they live in a country which is a treasure for the world ... but not yet.

TEN

Our cash killed Bedo

Sunday, July 2, 1989, Perinet.

Saturday night when the Bank arrived in Perinet he was in the bar of the Buffet de la Gare—where else to be available for tourists except Perinet's one hotel? (It has fifteen rooms, and doubles as the railway station.) Would he take the World Bank people out Saturday night in the rain? Of course. He sat with a bottle of beer, playing dominoes in English with two elderly ladies until we were ready—a 19-year-old professional guide, working on his English over the domino game.

I gave him the present I'd brought—Stephen Nash's portrait of thirty kinds of lemur saying *Arovy izahay*, 'Protect us!' I also had one for the hotelkeeper, Josef Andrianajaka. 'How pretty!' Josef exclaimed, 'I'll put it up and the tourists will be delighted!' But Bedo sat studying it, reading off to himself the animals he knew.

He took a few of the Bank group out that night, spotting a tenrec in the undergrowth. He took the elderly ladies out at 6 next morning, Sunday, even though we had booked him for another Bank tour, doubling his profit for the day. He escorted the women back and caught up with us in the rain-spattered woods.

I first heard of Bedo four years ago in 1985, standing in a windy airplane hanger waiting for a dawn flight with

Amoco's Twin Otter to the stone pinnacles of the Tsingy. Photographer Frans Lanting paced up and down, saying 'I wish Bedo had come. There are two great naturalists in Madagascar. One is Georges Randrianasolo, and one is a 14-year old boy named Josef Bedo. I so wanted to take them out to the Tsingy together, and to see that boy realize there are other people in the world who think like him.'

At 19 he was a dandy—orange-tan leather jacket, suede desert boots, and a reddish umbrella to keep the rain off his finery when not actually plunging through the woods. But he did plunge through, enthusiasm undimmed by rain, the drizzle in little drops perched on his bushy black hair. In his caramel-colored jacket and red umbrella he led our party of World Bank people through the Perinet rainforest. The Bank group peered up into the leafy sky, with grey-white cloud above and faces full of rain. They kept saying 'Where? Where? Where?' as Bedo told them which tree-trunk to locate, which branch to follow out, and then, see, in the fork below that bunch of epiphytes, two wet, immobile, nearly invisible indri—a trivial task for him, an eye-opener for the Bank mission.

He died because of jealousy. We, the foreigners, were responsible.

❦

Bedo was the son of the chief forester at Perinet. The tiny natural reserve and larger forestry plantation were overseen by Bedo senior, who lived in a two-room wooden house and scrupulously kept the threadbare uniform issued to him ten years before. Little Bedo was a natural naturalist. He knew where the tenrecs had their burrows, what hid in hollow logs, where couas nested and where the indri slept. He was the one who discovered that aye-ayes live at Perinet where no one expected them, and claimed that they are more social than scientists' wisdom pretends. I've seen him look across a valley at a velvet tapestry of trees, and spot the branches move, and know that kind of movement meant *Lemur fulvus*, and patiently explain to those of us with binoculars how to see what he'd seen.

Frans Lanting, balked at taking him to the *tsingy*, signed him on as a field assistant on the next trips—to Berenty in the south and Ranomafana in the east. Frans paid for his secondary schooling and spoke of future prospects if he would try to train formally. Bedo's friend Hubert hosted him in Antananarivo, trying unsuccessfully to steer him away from the temptations of alcohol and pot.

Bedo worked hard at English. He dreamed of someday visiting the States, perhaps even the fabled Duke University. But school was hard to stick to. Guides from Tana would turn up and stop in the school: 'Hey, Bedo! Hop on the bus! We have tourists to tip you if you come with us for the day!' The tourists didn't know how to tip any guide, let alone a sharp-eyed, English-speaking teenage boy. They gave him 5,000, 10,000 francs—three dollars, six dollars, even more for a morning. More than his father ever made; far more than the jealous other guides who tried to succeed by mere dogged learning, without Bedo's passion for nature.

Frans published a four-page story about him in the *National Geographic* children's magazine. Bedo at 16, stocky, grinning as Berenty's ring-tails, bounded up a sapling toward him. Bedo with Frans's headlamp, wide-eyed, watching one of Perinet's foot-long, rococo stick insects poised beneath a twig. Tourists came to Perinet from Kansas and California and asked for Bedo.

Bernhard Meier, co-discoverer of the golden bamboo lemur, spent three weeks in Perinet with a German film team. Bernhard asked Bedo to sign on with them at 7,500 francs per day—very good money for his age, though it translates to only $7. Bedo refused. He boasted that he could get 50,000 francs a day from tourists. Even though Bernhard did not believe it, the boy could drink up 20,000 francs in an evening. Bernhard had come bringing an offer of a scholarship abroad in his briefcase. After three weeks no more was said. Bedo blew his chance.

He gave money to his family; he bought a punk hairdo and a radio and flashy clothes. He was like a teenage pop star: rich beyond imagining, and rich with no idea what his talents could lead to in a wider, adult world.

Bedo went drinking at the bar after the Bank party left. We left him there with his beer after his two well-tipped sorties on a Sunday morning. There was a fight—not a confused brawl but a stand-up fight against someone jealous of Bedo's wealth and boasting. To everyone's surprise Bedo won.

The man he'd defeated swore he'd get him.

He caught up with Bedo at the river. He hit the boy on the head with a big stone and threw him in the river. Or else he waited as Bedo tried to flee across the stream, and threw large rocks, which hit Bedo in the head. Or else he threw small pebbles. The only place Bedo could find to hide was under the bridge. There is a deep whirlpool or eddy there, where the water sucks in beneath the bridge.

Maybe a stone hit his head.

Maybe it was the whirlpool.

Maybe he was just drunk.

Anyway, he could not swim. He had always been afraid of water.

Or something else, because in Madagascar, every happening turns instantly into myth.

Divers from Fanalamanga found the body five days later.

ELEVEN

The Bank goes to the forest

Oct. 4, 1989, Antananarivo.

It is happening! If ever I write a book about how conservation happens, maybe this is the high point.

Of course we haven't gone to the bottom, yet—there is no, or hardly any, change at the level of the villager, who is the woman that counts at the bottom.

But we sure are going to the top.

Barber Conable, head of the World Bank, is coming to Madagascar. And he is coming to Madagascar to focus on the environment. He is personally invited by the president of Madagascar. He will fly around in a helicopter. If Barber Conable expresses interest in any natural reserve, let alone gets to one, we've won at the top, because every Malagasy politician will know it.

It is now October, so the fires will be burning, blown upward in the wind of their own making. Orange flames rise toward the crest of the ridge on this side, and spend their heat at the mountain top. The dark, brooding, cooling green of rainforest sometimes cloaks the downward slope beyond.

Oct. 5th, afternoon.

Barber Conable isn't just having a half-hour look down at a forest from a helicopter! He is actually setting foot in one!

The government is resuscitating the Micheline, the luxury two-ended, one-car train that used to zip up and down the railroad line with comfort for the elite. I haven't been in it since my parents docked in Tamatave in 1963, ready to take me home to get married, and my train-loving father and my mother and I rode it up to Tana past all those waterfalls on the escarpment. Barber Conable will travel to Perinet in the Micheline!

José Bronfmann, the World Bank rep, and I simultaneously clench fists and mutter 'The indri had better perform!' José goes on to say that their eerie wail is the most evocative sound he knows. After the indri do their stuff, they'll fly Conable back to Tana for his interview with the president.

José has planned it all. Conable will see his rainforest, no fear. Apparently José has never got over his very first field trip here, when little Bedo took José and his daughters out at night in Perinet. They saw this and that—a chameleon, an avahi. Then, not ten meters away, in a tree there crouched two aye-ayes. That was the first confirmed sighting of aye-ayes in the upland rainforest. (Not counting Bedo, who made all the 'unconfirmed' sightings. It was Bedo who found them that night, too.)

Aye-ayes—in Perinet—with one's daughters visiting the new post—could have been the omen that would usher in the World Bank's environmental road to Damascus. Come to think, if St. Paul had seen an aye-aye we could all have been environmentalists for 2,000 years already.

So José wants to pass the benediction on to the Conables. Mrs. Conable, who is said to have fostered all this environmental awareness in the Bank, is also coming.

Wow!...

October 26, Antananarivo.

I am worried about the education sector of the Environmental Action Plan. Unesco is paying me to follow up. But the French will fund culture, not environment. The Americans will focus on biodiversity in a few particular areas. German Technical Aid, my best hope, has a big fat credit of 10 million Deutschmark voted—but for a loan. The Malagasy refused. No more loans. Not for the environment. Not even for environmental education. So this part of the EAP is lost, unless the Bank itself puts it into its vast new education loan. But the Malagasy do have a point about no more debt.

This Thursday morning I paid my courtesy call on Léon Rajaobelina. He is the ambassador who came to dinner in New York four years ago with Richard and me and Tom Lovejoy, to grumble in our living room about national debt and iron-visaged structural adjustment and the new hope of debt swaps for nature or for development. He told us about Madagascar's brief thought of not paying its debt interest. The Bank said, 'You are an island. You'd soon have no ships calling and no airline. That's your choice.' (South Africa or Brazil can refuse to pay their debts, because international financiers shiver in their boots at the thought. Madagascar can't.) Léon chaired the meeting at the Bank last spring on the Environmental Action Plan, ending up striding out to take that phone call of congratulations from his president.

Now he is no longer ambassador. He is the new minister of finance.

I waited my turn in the inevitable unprepossessing waiting room where even the concrete seems faded, at the top of three flights of concrete stairs. I was ushered at last into the grand ministerial office with its mahogany desk and leather armchairs in a strangely hybrid style.

'Alison! I didn't realize it was you! Sit down here!'

We talked of Russ Mittermeier, in his new guise of president of Conservation International. We spoke of the Environmental Action Plan, grown from Russ's first draft to one of the weightiest policy questions for Madagascar, my environmental education bit being only a tiny fraction. We talked of the imminent visit of Barber Conable. Rajaobelina expressed doubts that he should stay in the Perinet Hotel de la Gare.

'But José Bronfmann wants him to hear the indri wail!' I protested. 'And, anyway, he knows what the Third World is like! And even more so, since he says the environment is his first priority in Madagascar!'

'Still, the president of the World Bank... Can we really put him in the Perinet Hotel? It is not exactly what he's used to.'

I bit my tongue not to say 'He'll be all right if he doesn't keep food in his room. Eleanor Sterling did that and the rats trampolined off her toes onto the table to reach her biscuits.'

'I think he'll manage just fine,' I declared.

The minister's cigarette reached its end. We both identified a courtesy call as lasting for one cigarette. We stood, shook hands warmly, and I bounced rather dizzily out into the hall.

As I reached the stairs, the minister of finance called after me, 'Keep the faith!'

29 October, Antananarivo.

Things to tell Conable: The government, out of cash, told all the civil servants in Tananarive they would be paid a week late last month. This came up in a conversation about shopping with my Berenty research friend Hanta Rasamimanana. She said 'If you went to the market three weeks ago, you could have had crocodile handbags which usually sell for 24,000 ($18) knocked down to 8,000 ($6); you see, that was the week no one in the city had any money.'

'What do you mean?'

'The government postponed our salaries a week. You can imagine the uproar that would make in France or America. Here people just said 'Oh, then ... we'll have to wait.'

Another thing: the Ramahatras' son Mamy has written that his Canadian university scholarship only covers tuition, not living expenses. He's not allowed to get a job. He figures he needs $750 a month to live in the dormitory.

His mother gave a great cry and couldn't even finish reading the letter. She worked it out, with all the zeros. One month for Mamy equals four months of her salary and her husband's added together. They have already sold their car to pay for his airfare. She dreamed all night about serpents and crocodiles, convinced that Many would have to come home after he had worked so hard to leave.

This, mind you, is the president of Tananarive province and his wife—like a state governor, of the most important of Madagascar's six states. They live in a large official house with garden and view, I suppose half a dozen servants, a car and chauffeur for state (but not personal) business, and hospitality with the best of Malagasy cooking, elegantly served, on a long table with white embroidered cloth. The family of well-behaved children join family guests and official guests for every meal. I wish we lived so well.

But translate that lifestyle into dollars—and it vanishes. You might as well ask them to walk out in their garden and jump to the moon as to pay for a student son in Canada. Of course, the real problem is that they're honest, so there's no Swiss bank account.

And these are the people, from provincial president down to the market handbag-seller, that the Bank asks to 'adjust' their economy to demand less.

So will the Bank itself come through with aid, not loans?

Maybe the loans will just backfire. There's a Malagasy proverb:

Don't get lucky like the chicken.
It's carried on the head,
But only going to market.

November 9, 1989. Hilton Hotel, Antananarivo.

José Bronfmann
Resident Representative of the World Bank
for Reconstruction and Development
(The World Bank)
Has the honor to invite you
To a reception at the Hilton Hotel for
The President of the World Bank and Mrs. Barber Conable
7 p.m., November 9th, 1989
Ballroom, Second Floor

I lined up in the long queue of guests waiting to be received, stretching back through the balconied hallway to the elevator. The fashion, as always, is Parisian, with a predominance of *décolleté* black or white cocktail dresses and white high heels. Dresses and jewels show well in this crowd with all the men in dark suits and all the skin dark and lustrous except for the odd pallid diplomat. There are also a few West African diplomats in white flowing robes and blacker skin.

I think I am got up all right. I have on my best Nigerian long cotton purple-and-turquoise tie-dye with a breastplate of swirly turquoise and cocoa-brown embroidery. Malagasy silver bracelets and American Indian silver and turquoise earrings. Pretty good, considering I can also wear this dress on the beach if there are too many mosquitoes. Since I'm a head taller than most Malagasy, I'll be noticed. Or is this just nervous defiance at meeting the head of the World Bank?

A ripple of agitation sweeps the line. A phalanx of ministers and their wives sweep by, shaking hands with favored guests as they pass. At least I know the drill. The last reception I attended here was for Richard and me, with the Unicef representative making the introductions. The last line this long I waited in was at the UN for Premier and Mrs. Gorbachev, and the ripple and murmur of the passing phalanx was for the Nixons and Kissinger.

To tell the truth, the difference is not where I'm standing in line. The difference is that I care so much about this funny country, which isn't even my own.

And of all the people in the world, perhaps the president of the World Bank has most influence over the survival of the tropics as a whole. Not Bush, not Gorbachev, not the president of any one of the tropical countries themselves. Not even the secretary general of the UN, but the man with his hands on the biggest knot of economic threads in the international web.

OK, shuffle forward. The ballroom doorway looms.

It's a four-person reception line. José Bronfmann says, 'Hello Alison! Mr. Conable, may I present Mrs. Richard Jolly?'

'Oh yes! I know your husband well. In fact I was at the ACCSC meeting with him just three weeks ago.' Barber Conable is tall, balding, dignified, with a winning smile. He has also perfected the reception line handshake, a flowing motion from right to left that keeps the traffic going.

Charlotte Conable stops it. 'You wrote those articles in the *National Geographic*!'

I grin gratefully, astonished that she picked up on the name.

And I'm through. Waiters with trays of drinks, buffet tables with food invisible behind the swarms of gobbling guests, and two hundred people I don't know. The Conables stop to say hello as they start their circuit of the room, so I say I'll be in Perinet with them at the weekend. Charlotte

Conable looks pleased, gathers her red plaid mohair stole around her (it's gorgeous) and carries on to the next person.

In two days' time José Bronfmann will conduct his boss, with half the country's ministers, into the eastern rainforest at Perinet to talk about saving Madagascar's environment.

Saturday, November 11, 1989. Perinet, I mean Andasibe.

Vololomihaja Rakotomanana of the EAP support cell (she is called Haja) and I met at the railway station at 5:30 a.m., clutching our first-class tickets to Perinet. (I must learn to call it Andasibe, now.)

We trundled off past canals and shanties, with the first two stops still in sight of the Queen's Palace. The train went TOOT-TOOT like Mr. Toad. At one particularly long stop it tooted loud and long until a distant tootle answered like a mating call—a train coming the other way, to cross where we waited on a double bit of the single track line. Still among the green rice fields of the plateau, I dozed off.

I woke convinced I was still dreaming. The Malagasy facing me was reading the French edition of *Adjustment with a Human Face*. He was even intently studying the tables. He turned out to be an economist in the Ministry of Planning. When I said my husband wrote the book, he was wildly pleased, and sent *hommages* and admiration to Richard. Haja, giggling, confessed she'd been watching him read and me sleep for the last twenty minutes, just waiting for us to connect.[1]

The train reached the gorge top and started down the first step of the escarpment. The twisted white trunks of Madagascar's mid-altitude rainforest rose around us. It was spring, so the dark green forest was dotted with the yellow

1. G.A. Cornia, R. Jolly et al. *Adjustment with a Human Face* (1987).

and russet of new leaves, and a tree in fluorescent rose-madder flower.[2] We inched down the gorge, hugging the folded tree-clad walls, while the river hurtled from boulder to boulder below. Then a brief glimpse of a dam, a quick dark tunnel, and we were out on the face of one of the two monolithic granite mountains which are the gateway to the Mangoro Valley. The valley is a north–south graben, a rift valley between two faults. It runs all the way north to Lake Alaotra, the rice-bowl irrigation scheme which is being so eroded.

Anyway, the little river we had been following was now at least five hundred feet below us, having mostly got there through turbines that provide a large share of the country's electricity.

We straightened out and crossed the valley floor for another hour, past the town of Moramanga. (The Conables' helicopter will take them as far as Moramanga before they pick up the Micheline.) And up again through more rain-forest to Andasibe.

At last Haja and I got out into the Buffet de la Gare.
M. Joseph Andrianajaka, the proprietor, bustled up to us beaming. He was still beaming when we said that we, and in fact the whole support cell for the EAP, were planning to stay overnight. He has only nine rooms in the hotel upstairs and six brand new thatched bungalows, triangular wooden structures in a newly cut niche in the laterite hill. Joseph's bungalows are so new that no one has actually flushed the toilets yet. No wonder he's worried.

I must take a picture of the cut in the hillside behind: 1-inch-deep forest topsoil, 2-foot-deep eucalyptus roots, and solid sterile laterite from there apparently all the way down

2. *Tristemma orientalis, Melastomataceae.*

to the middle of the earth... This must be the very slope where my professor John Buettner-Janusch saw ruffed lemurs in the 1950s. He'd risen to take a leak at four in the morning, peered out the hotel toilet window. There were a whole family of the deep-furred black-and-white animals. John rushed back for his camera, ran back to the loo—and the camera jammed. He ran to wake Dr. Paulien, director of the Institut Pasteur, who then woke up his son, who fetched his own camera. They all three stood on the toilet and on each other to peer out of the window. Dr. Paulien's son dropped his camera in the toilet. Three witnesses, no pictures. So far as I know no one has seen ruffed lemurs here since.

This is also where I saw the golden-limbed diademed sifaka soaring against a blue-black storm cloud on my very first trip to my very first forest. They've all been eaten, like the ruffed lemurs. Indri, which are taboo, remain. If the proposed World Bank-funded National Park of Mantadia becomes a fact, it will hold 10,000 hectares of forest just beside here where all these creatures still live.

We had lunch with Eugène Rasolonirina, called Solo, little Bedo's best friend. To hear Solo tell it, Bedo never touched a beer, never smoked a reefer, never provoked a quarrel. The body was found with its arms still up as though protecting his face from the flung stones. (As though after five days in a flowing stream the body would be in the position it died.)

Never mind. Little Bedo is now a myth and a symbol, here and abroad. Frans Lanting wants to establish a Josef Bedo memorial scholarship for young Malagasy naturalists. Perhaps the first should be Bedo's beautiful sister, Eugénie. She has made it to university and through the first year exam where only a third of the students pass. Now she is back on vacation in Andasibe, tramping around in a pretty yellow shirt and rubber gumboots, showing a passel of German naturalists where to find caterpillar food plants for their studies of rainforest moths. Let Eugénie carry on the hopes of her family, and may little Bedo rest in peace. A cascade of white datura flowers hangs down as though in mourning by the bridge where he died.

Saturday evening. The cortège is arriving!

No, it's only the EAP support cell. Guy Razafindralambo and Levy drive up in the cell's Land Rover. Viviane Ralimanga arrives in another with a television set for tonight's show. With Haja and me, the whole cell is here now. I'm now the team naturalist: definitely technician status, not official. Almost no officials are coming except ministers.

The cortège roars into town. About twenty cars. Not ministers. It's bodyguards, local officials, extras. The forest guards all stand to attention in a row, then relax. They are wearing new khaki uniforms with green and sky-blue epaulets: the first new uniforms in ten years.

The sky turned from sun at noon to thunderheads at three to black with flashes of sullen lightning by the time the cortège of officials finally rolled up at five. 'If only the heavens don't open for the grand arrival,' I thought, with crossed fingers.

At six there was a toot-toot! up the track. The Micheline, a one-car train with airplane seats, came into view, headlights set close together like slightly crossed eyes. It drew up at the station archway bearing its cargo of dignitaries. The heavens opened.

All the little children who'd been waiting on the other side of the tracks squealed and dived into sheds for cover. Barber Conable held a black umbrella, one presidential arm sticking out of the Micheline doorway, to shelter his wife. Josef the hotel keeper shoved his 7-year-old daughter out into the rain in her white organdy dress with an armful of pink orchid sprays for Mrs. Conable. (Orchids grow here.) Journalists pushed forward to photograph the child; the rest of us pushed back under the station arcade. Those with more sense of protocol than self-preservation pushed forward again.

In short, it was just the kind of arrival you'd expect in Madagascar.

When all the damp dignitaries and ministers got out of the train, upstairs to relieve themselves, and down again to the great main room, it turned out they all had to go out again to visit a rural health center. They were two hours late, but the people had been waiting all day, and the minister of health was along, and so...

I watched them leave, thinking sooner them than me. I've done my share of visiting health centers. Levy and I settled down to a bottle of fizzy water, thinking wistfully of beer, but it was going to be a long evening. I told Levy about an impromptu visit I once made to the Perinet maternity post, which only had mattresses for half the women. The rest brought their own blankets and lay on the floor. An average baby here weighs about 2 kilos, 4.4 pounds. Of course they were beautiful: minute, doll-like newborns in knitted caps, and their mothers with that beaming glow that comes on the faces of new mothers anywhere in the world.

Levy said people in Tananarive know even less than I do, and that they would be shocked in a rural maternity post. Come to think, he's only young and thinking of marriage, so he's probably shocked at obstetrics in general.

Everyone came back. We settled to business. Most of the station hotel is a single room, a restaurant with long tables and white tablecloths. Round the walls are panels of

Malagasy scenes and wildlife posters, including Stephen Nash's lovely joint portrait of the lemurs that I brought two years ago, *Arovy izahay!* The hardwood floors are polished every morning with half a coconut shell gripped by naked toes. Joseph, the proprietor, and I remember this hotel in the days when he gave you the handle to the only loo instead of a key. Now it is smart, painted, prosperous. Joseph has put a triumphal arch of green palm fronds over each doorway The pre-World War I arches and gables of its construction lend themselves to pomp, even though it is only a railway station.

The pillars that support the beams are rosewood, a foot wide on each face, ten feet tall. They don't grow them like that any more. Josef says there are no rosewood trees that size within perhaps 70 km from here today. Maybe the Conables can admire the architecture when bored with the talks.

Maxime Zafera, the new minister of Water and Forests, gave the introduction. Then Philemon Randriamanarijaona, director of Water and Forests, described the forest estate of Madagascar, gesturing to a map as tall as himself. Guy Razafindralambo, head of the support cell, talked longest. He outlined all of the Environmental Action Plan, and the hoped-for fifteen years of World Bank loans for the environment. He said that they have been preparing the agreement for two years already and will need one more. Three five-year loans will take them from 1990 to 2005. Guy actually thinks he will have changed the nation by then. It seems a fitting goal.

Guy showed two videos: one of the slums and sewers of Tananarive (many of the town's alleys are human dung-heaps) and one of Lake Alaotra. The camera flew over the bleeding lavakas of the hinterlands and waded through silt slides in the paddy fields. It wasn't technically proficient, being done for Malagasy broadcasting. Viviane sat with a fast-forward control trying to find the bits Guy was talking about. I wondered if Conable, who must have had hundreds of videos specially prepared for him, appreciated the effort needed to produce even one multi-purpose tape here. Or that Viviane's brother specially drove the TV set and VCR all the way here from Antananarivo so there would be one.

My eyelids drooped, and I'd not done a third of the Conable's schedule today. Barber Conable permitted himself to remark, in English, 'If one picture saves a thousand words, just think how many words we've saved!'

Lights, mercifully. Dinner at last. *Potage de legumes*. The pungent local fast-stream crayfish with garlic butter. Goose, no less, with many vegetables. Hotelkeeper Josef outdid himself, and rose to the occasion in triumph.

Good night, good night. The Bank still in confusion who gets what room. The Cellule kept our toehold on a bungalow. I shared the double bed with Haja. Levy got the single. Guy could have had Levy's bed or floor space but opted for the car. Viviane and her brother drove the television all the way up the rain-slick road to Tana.

Good night.

Sunday, Nov. 12, 1989. Andasibe.

Good morning. Haja and Levy rose briskly at 4.30 a.m. I resisted for another 15 minutes, but the whole world was chattering outside our thatched walls.

Breakfast with the Conables and Minister Zafera of Eaux et Forêts. Even at breakfast, I said that Perinet's charm is the richness of its fauna. It's a mid-altitude forest, so it does not have large trees or high canopy. Instead it has solidly braided undergrowth, which gives layer on layer of foliage for different kinds of mammals, birds, insects. (Martin Nicoll first explained this to me, talking about his tenrecs, but it explains why naturalists work here year after year without even beginning to know everything. Strange that the majestic lowland forest of the Masoala Peninsula is actually less rich in mammal species than the wiry woods up here.)

I hope I explained it quicker than I've written this. At breakfast all they really wanted was a transfusion of locally grown coffee, not more talk. Mr. Conable kept interrupting me and making jokes, as though people usually listened to him. There hadn't been rats in the night, thank God. There had, though, been freight trains practically in the bedroom all night. And one of Josef's pigeons flew into the bathroom and banged about until presidentially evicted.

I asked Charlotte Conable if she liked butterflies. She said, 'Oh, yes!' and Mr. Conable began to describe American polyphemus moths with great eyespots on their underwings. That settled the first stop. Levy and I scooted up the road to warn the German naturalists. The cortège set off promptly at 6 a.m., with the first stop by the German tents.

Professor Wasserthal of Erlangen–Nürnberg University stepped forward, trim in khaki field clothes and a silken neck-scarf. The camp looked just as a field camp should: four dome tents, a one-board table, a path to the black-water stream with cooking pots drying by the water. By far the biggest structure was the screen-cloth butterfly house. Professor Wasserthal actually slept inside to guard his precious cocoons against mice and mouse lemurs.

I babbled to the professor that we had about three minutes for this unscheduled stop, and to the others that these were this month's representatives of generations of Perinet naturalists.

The Conables were ushered into the cloth house. Half a dozen species of sphinx moths slept on the walls—tiger-striped, grey-bark color, large, small. Professor Wasserthal does research on the co-evolution of moths and orchids. He showed them the most famous sphinx of all: brown, about 5 inches long, with rose-pink furry body and a proboscis curled round and round and round like the spring on my grandfather's gold watch.

This is *Xanthopan morgana praedicta*: the moth predicted by Charles Darwin. When Darwin first saw Madagascar's comet orchid (*Angraeceum sesquipedale*) with its 20 centimeter nectar spur, he predicted that there would be a sphinx moth with a tongue 20 centimeters long to pollinate the flower. The hand-sized flower waits, waxy and intact, for a month or more until the right sphinx moth finds it, and some are never pollinated at all. (This, of course, is why orchids bloom so long in Western sitting rooms and on the front porches of Malagasy shacks: they are waiting for their moths.)

The most spectacular creature outdid even *praedicta*. Last night Professor Wasserthal caught one of the largest moths in the world, the Madagascar comet or luna moth, *Argema mittrei*. The size of a soup plate, greenish-yellow, long-tailed, with black and red eye-spots on its wings. It slept on a cheesecloth drum the professor brought outside for everyone to admire. He explained that it sleeps high in the trees. If disturbed in the daytime it lets itself fall, twisting and floating in the air like a falling leaf.

The visit took six minutes, not three, because the Conables kept asking questions. Then it turned out that the Conables had each been trustees of Cornell University! Charlotte Conable has written a book titled *Women at Cornell: The Myth of Equal Education*, giving the other side of my father's *A History of Cornell*.[3] (I feel very stupid not to have known that, but at least I have a good read to look forward to.) And Barber Conable spoke of mother's

---

3. M.G. Bishop, *A History of Cornell* (1962); C.W. Conable, *Women at Cornell: The Myth of Equal Education* (1977).

paintings, and then started quoting Pop's limericks from the *New Yorker.* Conable said, 'I can't believe this! To think of meeting in Madagascar instead of Ithaca.'

Another few hundred meters, to where Minister Zafera and all the Eaux et Forêts guards were waiting. Their show-and-tell was even better. They'd lined up three of the local chameleon species on little trees. The thumb-nail size Brookesia was good. The foot-long Parson's chameleon was a star. The male elected to go luminous green, and to shoot out a tongue longer than its body at proffered grasshoppers. They had a tree boa too (*Sanzinia madagascariensis*). As usual the Tananarive Malagasy were horrified—they reached out tentatively and then jumped back without touching the snake. As usual the foresters explained in vain that there are no poisonous snakes in Madagascar.

The best was last: a *Hemicentetes*, the little gold-and-black-streaked tenrec. It was released from its basket and ran about wriggling its hedgehog nose, bewildered and cute. The Conables liked hearing that it makes supersonic calls to its babies by rubbing the quills on its back. The little creature ended by running between Mr. Conable's feet, cowering between the two sheltering rocks that were his big black shoes. Then it escaped over one of his toes, and one of Mrs Conable's, before finally scuttling off to the bushes. I have a photo to prove it.

The Conables bravely plunged into the forest, following a forester who claimed to see branches move in the distance. They found the indri—and for them the indri sang. This is why José Bronfmann engineered this whole trip to Perinet: so that the Conables would hear the indri's magic song.

We drove off between hillsides black with new tavy clearings, smoke still curling up from the tree stumps. After last night's rain we hadn't even the drama of orange fire on the hills, just the black slopes of the forests' grave. At Beforona, island of sense and sanity, Jean-Louis Rakotomanana, the saintly agronomist, showed us round just as he showed the World Bank mission around last July. In honor of the Conables he wore shoes. His English is almost as magisterial as his French: perhaps some people have the gift of tongues, like being born Mozart. His visual aids included a chart on soil loss from sloping *tavy* fields: 500 tonnes of topsoil per hectare per year.

As we left, Guy Razafindralambo said to me, 'What we must do is spread the lessons of Beforona to all Eastern Madagascar,' while Barber Conable simultaneously said to José Bronfmann, 'How can we spread the lessons of this tiny experimental farm to every peasant.'

'You have the chance,' said José. 'You are off to see the president of Madagascar in an hour's time. You tell him that only political leadership and will can save this country's environment.'

This is Madagascar. The helicopter was late. No, it couldn't even land in the grey cloud of the rainforest zone. No, there was no way to be sure, because the telephone hadn't worked for the last three weeks.

'I guess I shouldn't care if I'm late for Arap Moi,' remarked Mr. Conable. I hooted with laughter, supposing he'd substituted Kenya's president for Madagascar's as a high-level joke. It seems, though, that following his session with President Ratsiraka he flies to Nairobi and straight on upcountry to Baringo District with President Moi.

The day which began with a comet moth, a hedgehog tenrec and the song of the indri ends with two successive country presidents.

The helicopter eventually did whisk him away.

I drove back to Tana with Guy Razafindralambo. Guy waved at the *tavy* fields scraped back almost to the ridge crests.

'Look at that! Look at that landscape!' said Guy. 'You understand, in Madagascar we must succeed. We haven't a choice. It is not like Europe here, or America. If you desertify part of your country, you have enough to spare. Even in Europe, if France were destroyed the Frenchmen would just go off to Germany or Switzerland.

'But here we are forced to succeed. We have no choice. If we destroy Madagascar's environment and turn our island into a desert, what will we do?

'We'd have to swim!'

PART III

Environment and development

Putting conservation into action brought its own problems, exemplified by the Ranomafana National Park. The biodiversity arm of the National Environmental Action Plan, the NEAP, focused on a few great Integrated Conservation and Development Programs. These ICDPs built on the Biosphere Reserve idea pioneered by Roland Albignac at the aye-ayes' home, Mananara. A central core would be off-limits to anything but research. A peripheral zone would ensure conservation of the center, open to all local non-destructive use, but not to *tavy*. Outside this again would be help for the fringing villagers to alter and improve their lives. Two NGOs or universities would contract to carry out the plan: one for research on the natural ecosystems, one to bolster local conservation and development. The donors had been bashed for their mistakes elsewhere. In Madagascar, let the NGOs get it right.

All clear on paper, but so naive.

First, local people had used the whole area of the reserves for generations. The integral cores were long since protected by French fiat, but villagers knew where the ancestral tombs lay inside the parks. Second, *tavy* was and is a principal use of the forest. Third, even the comparable cost of wood, herbs and crayfish was a challenge. In Ranomafana most lemur meat was *fady*, taboo, but in the Makira region by Masoala a recent study found that 57 percent of family cash income is from bushmeat. All told, returns from the forest are far higher than the outsiders guessed at first: in the Mantadia area near Andsibe it was way over $100 per family in a place where national per capita income was around $220, averaged between rich and poor—and these families were poor.[1]

Fourth, most important of all, the geography was wrong. Each vast forest tract had a 5 km fringe of villages to be 'improved.' 5 km is nothing to people who walk that far to drop in on a neighbor, sometimes 20 km to sell a chicken or a basket of fruit. Genese Sodikoff, writing about the Mananara Park, gives a wonderful riff on the importance of feet. The forest guardians that she knew measured their prowess by the strength of their feet. Feet with strong calluses against thorns and sharp stones. Feet with strong toes to curl into slippery clay and over rainforest roots. Feet that can walk for days over mountain trails. An arbitrary limit of 5 km between villages to help and those to leave aside could only send ill feeling, even though there might be far too many villages to help even with this criterion.[2]

On top of all, the first evaluations would come within three years. The NGOs had to show results. In short, an impossibility.

Ranomafana National Park faced the same dilemmas as all the other ICDPs. It is the outstanding success of Ranomafana's research that brings the resentment by surrounding villagers into sharp relief. Chapter 12 describes my first visit to

1. P. Shyamsundar and R.A. Kramer, 'Tropical forest protection: an empirical analysis of the costs born by local people' (1996); P. Shyamsundar and R.A. Kramer, 'Biodiversity conservation—at what cost? A study of households in the vicinity of Mantadia National Park' (1997); C.D. Golden, M.J. Bonds et al., 'Economic valuation of subsistence harvest of wildlife in Madagascar' (2013).

2. G.M. Sodikoff, *Forest and Labor in Madagascar: From Colonial Concession to Global Biosphere* (2012).

Ranomafana, in 1986, when a new species of lemur had just been discovered: a red bamboo-eating creature with golden chipmunk cheeks—nobody yet was sure what it was. By 1987 the indefatigably enthusiastic Patricia Wright allied with Malagasy conservationists to dedicate the golden bamboo lemur's forest as a future national park. Chapter 13 gives an account of this period—already with a blooming lemur research camp, but the problems of local people already looming on the horizon of the eager naturalists. Chapter 14, 1993, with the National Park established, tells how the king's spokesman of a nearby village stated bluntly that conservation is the new colonialism. The park simply took away their land. Integrated Conservation and Development programs had been lavishly funded by USAID, but Chapter 15 shows how the two parts of the Ranomafana program came spectacularly unstuck, an extreme instance of what was happening to such efforts all round the island-continent. The dilemma continued even in 2005, as we see in chapter 16, which ends with the DreamWorks visit.

Nonetheless, Ranomafana's research arm has since deservedly become one of the best-known successes of Madagascar. In Chapter 17, 1997, Patricia Wright and I meet the formidable president of Madagascar, Didier Ratsiraka: she to promote Ranomafana National Park, myself the forthcoming International Primatological Congress. A summary of current research is on the website of Stony Brook University, but a taste of researchers' joy is given in Chapter 18, where Malagasy and foreigners join in 1998 to celebrate the first international congress since 1985: joy summed up as 'Madame Berthe was Dancing.'[3]

3. www.stonybrook.edu/commcms/centre-valbio.

TWELVE

Golden bamboo lemurs of Ranomafana

Ranomafana lies on a narrow neck of eastern rainforest bisected by the main road that runs from the city of Fianarantsoa, high on the central plateau, all the way down to the coast. In its forest lives a red bamboo-chomping lemur with surprised golden eyebrows. Its discoverers didn't even know what it was.

In 1986 Patricia Wright of Duke University Primate Center attempted what seemed like a wild goose chase up and down the tracks of the eastern escarpment, brandishing a nineteenth-century color plate of a bright red animal labeled *Hapalemur simus*, the greater bamboo lemur. She asked all along the way if anyone had seen this possibly extinct animal?

Jean-Jacques Petter, Georges Peyrieras and Georges Randrianasolo actually found a few greater bamboo lemurs ten years before, in the small sacred forest of Kianjavato, a long way downhill from Ranomafana. Pat had never seen those animals, which did not look much like her picture. Russ Mittermeier's Primate Action Fund (supported by gracious Margot Marsh, who had fed dinner at her home to the Napoleon tour), gave Pat a tiny $5,000 to hunt the elusive *H. simus*. Russ remarked later, 'Best investment I ever made!'[4]

4. Russ Mittermeier to Alison Jolly, email, January 20, 2014.

After Pat searched in vain up and down and round the escarpment, on perilous tracks, bird guides Loret and Émile from Ranomafana said, 'Oh yes, if you cross the loggers' bridge over the Namorona River waterfall, there you find those red bamboo lemurs.'

Almost simultaneously, Bernhard Meier from Germany also crossed that bridge to start a long-term study. Rumors swirled: someone had rediscovered greater bamboo lemurs—except that these were red.

So, after the Alton Jones's trip, Russ Mittermeier and I absolutely had to travel to Ranomafana. Perhaps we could glimpse the mysterious new creature.

To clear up the confusion, there turn out to be three different bamboo lemurs. Common small gray ones, the *Hapalemur griseus* species group, nibble many kinds of bamboo leaves and reeds and occasionally people's paddy rice. The red form, called golden bamboo lemurs, *Hapalemur aureus*, feed mainly on cyanide-laced growing shoots of giant bamboo. The even rarer grey greater bamboo lemur, *Prolemur simus*, shreds the tough bases of giant bamboo into something like a fistful of dry spaghetti. The little red animals inspired the creation of one of Madagascar's most famous national parks, but in 1986 we couldn't foresee that.[5]

The big question was which scientist had a right to be there, and who had the right to claim discovery.

November, 1986. Russ and I arrived in Ranomafana to call on Bernhard Meier.

We commandeered Loret Rasabo, a young guide with sleepy eyes, to show us the way to Bernhard's house. We picked up Loret at his village, Ambatolahy, 'The Village of the Male-Stone,' which consists of twenty mud houses. A row of

---

5. *Hapalemur griseus* is now reclassified as five separate species living in different parts of Madagascar, but all closely related.

tree-ferns sculpted into flower pots lined the road in hope of sales; it's a wonder there are any tree-ferns left. A path went down—not quite straight down, maybe not even a 45-degree angle—to the Namarona river. There are only three places in about 30 kilometers where you can cross the Namorona's waterfalls and rapids as it cascades down the escarpment. The loggers' bridge was just three long trunks over the waterfall, with flat board steps. No need for handrails, why bother? Just don't look down. If you fell, you'd surface several kilometers downstream—or not at all. I have no sense of balance, and can boast of falling off log bridges into streams on every continent except Antarctica. Over the Namorona sheer terror kept me upright to the far bank.

Uphill on the other side, like a very steep staircase. Then down. Then up. Later, Bernhard swore his house is only 840 meters from the river. Distances just feel longer in the tangle of forest. His house is an open lean-to with a view out over a little valley of bamboo. High forest rises on the slope beyond. Neat shelves, neat plastic sheets under the travellers' palm frond roof, gear neatly stored in rubber bags. We'd heard so much good and so much bad of our host! He was sent out here to catch lemurs for Yves Rumpler's lab in Strasbourg. A house can tell you its maker is efficient, artistic and poor. But not if he is good or bad in the ethics of conservation … and there was no Bernhard.

Right—we'd see the forest. An afternoon of sun in rainforest isn't for wasting. Loret took us up a ridge, and down, and up the next ridge, and down. 'Ah—oh' he called. A voice answered—Rajerison Émile (called Rajery) stood in the woods, in blue wool pullover and red-and-green tartan cap. He was taking notes on a troop of red-bellied lemurs he'd been following all day. A troop of two, just the male and female. No, not the tame ones by the house—these were an unprovisioned group with a 10 hectare territory. Meanwhile Bernhard was following another study group for comparison, ones with a 100+ hectare range. Who was this Bernhard who could get village guides to do all-day-follows on their own?

We traipsed back, up down, up down to the stick house. No Bernhard. His cook was bubbling something over the fire-stones—apparently a whole neighborhood came to the stick house? A cook and three guides? But where was the host? Russ was already using his flashlight to see the trail as we approached. I didn't because I see trails better in half-light.

Should we start back to the road or wait? I opted for going while there was still a little twilight. But twilight doesn't last long in the tropics—after the first uphill from camp I switched on my light too—I'd only brought son Dickon's tiny 'bedside' one because this was just supposed to be a brief social call in daylight to meet Bernhard Meier.

Of course you always imagine what you'd do if lost in a forest like that. I lagged way behind the others, at intervals anyhow. I made up my mind back in the Nilgiri Hills in 1979 that the only comfortable way to travel with Russ Mittermeier is let him walk at his pace and me at mine, and catch up an hour or so later. So I amused myself thinking: if lost, I might have to spend the night, but even the rain isn't cold and there are no dangerous animals…

Then Loret put his hand on a scorpion. Only time I've heard a Malagasy say, when asked, *Ça va? Non, ça ne va pas.* The beast was on a tree trunk just the height you grab for support. Loret was stung on the little finger, but the poison filled the heel of his hand, then worked rapidly toward the elbow. Russ, who always has everything, dosed Loret with Bufferin from his pack. No, it isn't fatal, or even really paralyzing; just like being stung by a whole colony of wasps.

Russ claimed the Bufferin was strong, strong painkiller, and Loret felt better.

There was no help for putting hands on tree trunks. I've always progressed in steep woods with my hands, to move at all, like an orang-utan. So I lagged way behind again.

At which point Dickon's light went out. The batteries died suddenly. I hadn't imagined a night in the woods without light, unable even to see my feet. The others had reached the cascading river already and couldn't hear me shouting. I stood where I was and waited till finally Russ got worried and came back. It turns out he travels with four flashlights. This is the difference between a background in the Amazon and one in Madagascar.

There was another light, and a brief glimpse of flash-lit face, a sweet smile and brown beard. Bernhard, returning from his all-day-follow! *Guten Tag! Wilkommen! Wie geht's?* Would you come to dinner at the hotel with us? *Zehr gern! Und bleiben zie zu Hause mit Mir!*

Dinner in the Hotel Thermale of Ranomafana, which is Malagasy for hot water. Natural warm sulfur springs bubble into bathtubs in the nearby Bains Thermale; the little hotel is a relict of Frenchmen seeking the health cure. Over the regulation stack of plates involved in a Franco-Malagasy meal we sat trying to figure each other out. Fortunately for me the dinner conversation veered only between French, German and English, sentence by sentence, without having to follow Russ in Portuguese, Spanish or Sranan-Tongo. Bernhard did laugh at Russ's German, which had the phrases of a 7-year-old brought up in a German-speaking family.

Bernhard did a lovely double-take when he finally realized my name. He has *Lemur Behavior* on the bookshelf in the stick house, and said he'd thought I was about 60. He was so grateful that *the* Russ Mittermeier and *the* Alison Jolly had come to visit him. (I began to feel about 60.) He spilled over with eagerness and excitement. He talked about the red bamboo lemur, gene pools and restricted reserves and patchy distributions. He talked about the only five spots with giant bamboo in all the forest he had surveyed, and how the red bamboo lemurs ate almost nothing but giant bamboo, though nibbling at a few other species, so their distribution was patchiest of all. His thesis had been the collection, founding and genetic analysis of a breeding colony of slender loris from Sri Lanka. He has, in short, exactly the theoretical background needed to bridge from the problems of managing small zoo populations to the problems of managing wild populations that will, in the end, be very small.[6]

Then he spoke of the Perlekette—the chapelet of pearls. Madagascar's eastern forest needs not one reserve, or two, or three. It needs a reserve system—protected areas not islanded but strung together by the forested escarpment like pearls on a string. There's a theme for a book.

In short, the untrustworthy animal-catcher seemed to be a trustworthy, idealistic, hard-done-by student. We finally found the answer to Bernhard's contradictions. If he'd thought I was 60, I thought he was 26, so wistful and excited... He's 36 instead, a ten-year teacher of high-school biology, used to managing budgets for German nature reserves. His confidence with the guides, with radio tracking and data collection, and even the way he got his house built, is the confidence of a man only one year younger than Russ and an authoritative Teuton to boot. He was sent out to Ranomafana by the scientific community and then essentially abandoned alone in the woods—a 36-year-old

6. A. Jolly, *Lemur Behavior* (1966).

Hansel with no Gretel—but maybe he is more manic-depressive than that... In the meantime, Russ had started right in calling him *du*, whether because of student status or forest comradeship or just because Russ's German dates from his own childhood. I followed suit, not to be stiff and awkward. We both hiccuped a moment after Bernhard told us his age, then settled back to *du*.

Meanwhile I grinned at the searchlight of Russ's charm turned on to a vulnerable younger man. I've seen so many women dithering over how quickly to succumb. Insulated by my ten years' seniority and by my blatant monogamy, I have the delight of travelling with Russ and being real friends.

When Bernhard first arrived he came straight from the airport to Ranomafana. Then he went back and then spent ten whole days liberating his camping and scientific gear and schlepping it by truck back to his study site. He found Patricia Wright of Duke installed in his campsite, having hired his guides, and having found the red bamboo lemurs. Patricia did see them first, unless you count Loret of the sleepy eyes and Rajery in his tartan cap, who showed all the scientists where to look.

Bernhard, by his account, rushed up with open arms, hoping to join the Duke party. Patricia was sharply suspicious of this unknown German sent out to catch lemurs. And the suspicions have ricocheted round the world ever since.

A mother red bamboo lemur sat in a tree crotch, holding the tip of a giant bamboo spike like a dwarfed drum majorette. She'd chewed it off so the piece she held was perhaps 3 inches in diameter, and three times as long as her body. She addressed it like corn on the cob, sinking in upper canines (lemurs have rudimentary upper incisors), and stripping fibers loose with a ripping noise. Then she and her half-grown son settled to crunching the exposed undercoat. Rip, rip, rip, crunch, crunch, crunch.

They don't eat like lemurs, and they don't move like leaf eaters. Indriids loll among the branches, languidly stretching out a hand to nibble a leaf-tip like furry Madame Récamiers. The bamboo lemur was in a hurry, grabbing and gobbling, while the juvenile male bounded about like Russ fighting to set up his tripod on the steep slope. We all decided that bamboo shoots must be high-sugar food—the primate equivalent of Coke and chocolate bars.

Russ and I flashed away at mother and baby while the drizzle turned slowly to thunder and lightening. We did have a patch of sun later. Bernhard led us up and down another ridge to high pristine jungle—big trees, apparently never touched by loggers, that met over a little granite-bouldered stream. There was another stand of giant bamboo. Once you know the signs, you can always track bamboo lemurs: the bamboo top spears are gone, so the stems leaf out at the sides in little tufts. Some stems were full of great holes near the base—inconceivable that a primate can chew holes like a beaver, nipping through woody stems 6 inches in diameter. Giant bamboo grows to 30 meters tall. It is just one endemic species. Bamboo lemurs spurn the introduced Chinese bamboos. Their diet, and distribution, seem to be tied to a single plant.[7]

But the mystery of bamboo lemurs remains. I saw lemurs of both the first and second round. The ones Petter and Peyrieras and Randrianasolo caught from the Kianjavato coffee station in 1972 (halfway down the road toward the coast) were dirty yellowish grey, with tufty ears, very different from Patricia and Bernhard's glowing red ones

7. *Cephalostachyum vigueri.*

with the creamy-gold chipmunk cheeks. I wonder how different these red ones are?

For decades almost all the forest that drops down from Ranomafana to Kianjavato has been cleared. There's a fascinating suggestion that both lemurs and bamboo came originally from lower forest, in the land now bare. The president of Loret's village says all the giant bamboo within the remaining forest was planted by people, in patches near long-abandoned villages! And now even the remains, the top step of the escarpment, are under threat. Bernhard walked Russ all over the adjacent mountain to show him cleared *tavy* fields hacked out of the bamboo, and me round to see a 'road' where a local contractor, Paul Rarijaona, wants to bring chain saws and tractors to scrape the hills bare. We may be able to stop him—these hills are the watersheds of the Namorona river. The Namorona's famous cascades provide electricity for the towns of Manakara, Manajara and half Fianarantsoa. Rarijaona is a true villain—surely no Eaux et Forêts man would have given him the permit to fell these particular slopes out of mere stupidity. Money has changed hands somewhere, to let Rarijaona make a profit off the electricity company watershed, which now happens to be the only known home of the red bamboo lemur.

When we finally went to bed in the stick house, one final, unnoticed leech dropped off the crook of my elbow like a ripe grape. The others found some too; they didn't say where but sounded upset.

I lie in my down bag, on my magic self-inflating mat, looking through the house slats to rainforest leaves. Sunlight already. Wriggle erect enough to look round the partition and see a recumbent khaki lump under the working shelf, so Bernhard is still asleep too. No noise but coua calls and forest frogs and snores, which implies that Russ and the guides and the cook are also asleep in the other part of Bernhard's stick lean-to house, around the fire-stones. Russ has greatly increased the floor space in the 'living room' by slinging his Amazonian hammock from the house poles, which lets the guides extend underneath him.

No point in getting up!

I protest I'm not asleep when a fruity and familiar voice booms 'You lazy bastards! In bed at 7 a.m.! Oh, hullo Alison—I might have known you'd be wearing an upmarket nightgown even in these circumstances!' (Richard's Sudanese caftan).

Quentin Bloxam, the big herp curator from Jersey, blows into the house, followed by small, blond, tanned Lucienne Wilmé. I hadn't really registered her before, thinking of her as bird-man Olivier Langrand's spiky-haired girlfriend. It turns out she's a long-term birdwatcher on her own account. She knows the trails round Ranomafana as well as anyone. It was she and photographer Dominique Halleux who first trained Loret and Rajery.

'All right,' booms Quentin, 'where's breakfast? And bring on the lemurs!'

So, on cue red-bellied lemurs come and pose right on the stick terrace of Bernhard's stick house! The male, shyer, with his nose so black and his teardrops of white by the eyes so white they're together a real face mask. The female, and the 'baby male,' a juvenile already on his own, are round and red and fuzzy and apparently cuddle-able. The baby male's cream teardrops are just starting. The female sports a few cream-colored spots on the bridge of her nose like freckles. She's much sassier, like all lemur females. She eats from your hand and leads her male all round the forest, he marking branches after her with the top of his head. They're apparently always monogamous, and the young may leave before the next birth season—any day now. No wonder the female is so round.

Everyone whips out their cameras, again… Quentin admits he is even growing to like mammals as much as herps, as though this were a sentimental weakness of age… The red-bellied lemurs unsentimentally gobble banana slices.

We kit out with enough lens-power for a small observatory, hung round us in belts and vests and knapsacks. Lucienne Wilmé adds a variant—a Sony field recorder for bird calls, while Loret shoulders a meter-wide parabolic microphone that makes him look like a space invader. Then we clump off down the trail.

Up and down and up to the second ridge, where Rajery calls 'Ah-oh' and points upward.

Seven diademed sifaka loll in the sun. Seven and a baby. The baby is perhaps three months old, past the stringy, jughead stage of new sifaka, and now a perfect teddy bear. It bounces on its mother, and hop-kicks its way up the branch and scrambles back and plays with her tail and starts all over again, while she absent-mindedly gives him a few strokes of her tooth-comb when she can slow him down. The seven are arranged on two white branches, forks of a dead tree in front of blue sky. The mother and two friends groom on one branch, four more on the other. These are Milne-Edwards' subspecies, very different from the grey-and orange diademed sifaka of Perinet. Their chocolate fur gleams rich as Malagasy rosewood, their backs are sunlit cream-white against the gray-lichen white of the tree.[8]

One of the mother's friends leaps over to the group on the opposite branch. The second follows, and the mother, so the tree branch blossoms with an intricate tangle of seven lots of

8. Now a full species: Milne-Edward's sifaka, *Propithecus edwardsii.*

arms and legs and poking muzzles. Then the baby jumps over too, on top of the heap. The last straw.

The branch breaks. Sifaka rain down all over the place like tumbling acrobats. Some catch themselves on low trees, some hit ground with a thump and bounce back at once to stems a meter off the ground. The baby hits about 10 feet from us, then clings to a vertical stem, looking about.

As usual with sifaka, no sound—not even the baby gives the bubbly coo of contact, though its mother comes toward it and it hops rapidly to her back. The others just sit looking dazed...

And so do we. Not one of us had the presence of mind to take a picture as they fell, for all the massed telephoto power aimed at them sunning!

So endeth the story of the enchanted forest of Ranomafana. There will be sequels. Will Patricia and Bernhard and Russ save the forest from the evil exploiter? Will Rajery and Loret, meanwhile, remain true to the salary Russ is offering them, and save those trusting lemurs from being eaten while Bernhard goes back to Germany to write his thesis?

I suspect the next chapter gets written in Washington. The main characters will be the smooth, smooth minister of Eaux et Forêts, and pompous trustees of WWF, with Russ and me disguised in the background in necktie and nylons.[9]

And some time we'll all foregather at Ranomafana, all who can, and the ones who were there this time will point out to others, with parental pride, the young red bamboo lemur which learned how to eat giant bamboo, and the baby sifaka whose weight was the last straw that broke the branch on the day the diadems fell down.

---

9. See Chapter 7.

THIRTEEN

Patricia walked the boundaries

After the confusion about what the red bamboo lemur was and who found it first, all the Western discoverers agreed to publish together—a whole new species. The remaining choice was whether to name it red, *rufus*, or gold, *aureus*, for its golden cheeks—gold won.[1]

Patricia Wright campaigned for the Ranomafana Forest to become a new National Park. Joseph Andrianampianina carried the torch within the forestry department. He had tried fifteen years before to create such a park to protect the watershed for the dam on the Namorona river, which supplies electricity for the towns of Fianarantsoa and Manakara. The same logging interests blocked that first effort that were now opposing Pat Wright. Pat was sure she and Joseph Andrianampianina would win.

In 1987, after the tour of US zoos with the Malagasy delegation, I came back to Ranomafana—for once with Richard, and with Hubert, who was Frans Lanting's assistant. Now, a year after the red bamboo lemur's discovery, Pat tramped round the boundaries of her proposed park, coming 'home' to a newly fledged research camp. But the loggers were still there, and a mishap might destabilize the whole project…

1. B. Meier, R. Albignac et al., 'A new species of *Hapalemur* (primates) from South East Madagascar' (1987).

August 2, 1987, Ranomafana.

On the switchbacks down by the Ranomafana waterfall, a battered green Land Rover braked and disgorged Pat Wright, Patrick Daniels and about ten other people like the clown car at the circus. They were just back from walking the northern boundaries of the proposed National Park. Pat stood a-straddle, in great big rubber boots with a tread like truck tires, and her raincoat spread and her hair out to her shoulders, totally in charge of her team. She said they'd walked for a week, to villages days from a road. There was part of a road once but it has gone—other parts never had one. They stayed in five villages on the five nights, and visited about twenty. In each case they held a *kabary* about the new reserve, and the change, and help, that would be necessary for settled agriculture. Just as there was once a road, so there was once irrigation in many of the valleys—if it were working again there would be an alternative to slash-and-burn *tavy*.

Strangely they were welcomed by the villagers because last year the region was hit by a devastating cyclone. Many *tavy* fields became landslips, and many people were killed—not out working the fields, but women and children in houses that were carried away.

In each village after the discussion, they solemnly took polaroid pictures of the assembled elders for them to keep. Many had never before had their picture taken.

Pat's group retreated uphill to their camp. Richard and I went downhill to the Hotel des Thermes in what was once a French spa town. Rendezvous 8 tomorrow at the logging village.

August 3.

Which we were late for. Seduced by warm sulfur shower, or in Richard's case the warm sulfur swimming pool.

But now it's real life. Up past Ambatolahy, the village of the male stone, to the mud huts of the loggers. They smile and wave and seem to feel none of the anger we bear toward their bosses, and sometimes even to them. They get 1,000–1,300 FMG, 80 cents to $1, for each of the planks they carry out on their heads—cut, sawn and carried. I tried lifting one in the woods and could not raise it off the ground. As my friend Joelina Ratsirarson says, if you had any other way to make a living, wouldn't you take it? Some 350 people from Ambositra on the plateau work in these woods, only 30 of whom are local Tanala. The Ambositra men are far from their wives and families; the Tanala are guides; it wouldn't take much tourism or development to hire the 30 men who now carry logs.

We go down the path in the sun. Over the foaming rapids and falls of the Namorona by the loggers' bridge: three long tree trunks with flat boards nailed across at intervals, each board about the shape of a staircase step. The middle tree trunk has about given up since last year. Still no handrails, of course, with a medium-sized waterfall above and a very big one below. Shiver.

Up trails, under trees only 40–50 feet tall, but with all the richness of a… well, of a selectively logged forest. Pat mourns 'On our boundary walk we went through untouched woods with trees this big around—she holds out her circled arms with the fingertips 2 feet apart—and the lianas this big—the circle of her two hands with the thumb-tips opposite fingertips. 'Its a long, long time since I've seen lianas like that.'

Richard and I went to see the sawmills in the woods, lashed sapling trestles with a man above and a man below wielding the two-handled saw. Planks and wood chips everywhere. One tree lodged its canopy into its neighbor, so they left it, poised on its cut point and leaning over the trail. It was too thin to be worth cutting in the first place.

The woodcutters loped down the trail past us, those fiendishly heavy planks balanced on round rag cushions on their heads. They never failed to smile both warmly and politely as they passed us. The sound of axes rings continuously—toc, toc, toc—a tocsin, not a clock.

How Pat must suffer to see her study site mutilated. But she is right: there must be a long-term solution, not short-term antagonism toward these smiling, courageous men.

The social life of sifaka. Michael Todd from Duke and Bettina Grieser from Tübingen are each chasing a group of *Propithecus diadema edwardsi*. This week it's five-day samples of ranging, foraging and social behavior. They do this once a month, now standard primatological practice. The quirk is that every animal is a focal animal, so all its behavior, especially social interactions, gets recorded. This is OK for Michael with a group of one adult male and two females, given *diadema*'s normal indolence. Bettina instead has a group of seven: an adult male, an immigrant male, two females, a female yearling (the one who made them all fall down last year), a subadult male and a new baby. The young male and the infant are from one mother, the yearling from the other. It seems there is often a two-year birth interval. The two females don't like the immigrant male, who hangs out on the sidelines.

The males are even more submissive than the sifaka males at Berenty. They almost invariably wait until their females have finished feeding in a tree and move out before moving in. If they do try it there's a scuffle and a spat and they're chased off, so we mainly watched them hanging wistfully in trees watching the females eat. Pat figures it's a matter of crown diameter. The rainforest crowns are squashed together, smallish, with few fruit or new leaves. The females need first go.

The immigrant male is odd. He started in Group II, the big group. Just after the babies were born he moved to little Group I. The male welcomed him. Then, though, there were baby's screams, and the observers rushed up to find the infant wounded and dying. Infanticide in sifaka???[2]

The females avoided him, both before and after the infants' disappearance, but the male continued to sit with him and groom. We've lots to learn, still, but my guess is that (1) two males have more clout defending territory than one, and (2) a male who spends most of the day sitting all by himself in a tree while his females stuff themselves—gets lonesome.

Bettina is with the big group where the young males have been playing. They are the most teddy bear of sifaka. They bounce usually no more than 10 feet at a go between the close-set vertical trunks of the rainforest: bounce bounce bounce bounce like a chocolate plush tennis ball. Pat, who loves them, claim they must know they are beautiful when they stretch out their long limbs on a high white branch in the sunlight, or pose framed before the unfolding view of rainforest ridges.

We found them on a trail, then plunged down and up a ravine as they bounced effortlessly across at their own level. Treetop locomotion is supposed to be no more energy-efficient than ground locomotion. I suspect this is the theorizing of laboratory scientists who think the earth is flat. It isn't—it's downhill on slippery clay and frequently on one's bottom, and uphill with backsliding on more of that damned clay and thickets and roots and nice little lianas strung across about 1.5 feet off the ground like naughty

2. Infanticide by unrelated males turned out to be a major source of mortality in these sifaka. P.M. Wright, 'Demography and life history of free-ranging *Propithecus diadema edwardsi* at Ranomafana National Park, Madagascar' (1995); S.T. Pochron, W.T. Tucker et al., 'Demography, life history and social structure of *Propithecus diadema edwardsi* from 1986–2000 in Ranomafana National Park, Madagascar' (2004).

children doing it on purpose. Of course you get fitter as well as more observant after a week of it—but neither Malagasy nor for that matter wild primates get fat. Nor do field primatologists. Pat got out her Swiss army knife and fixed us all lunch: a smallish hunk of bread and cheese and an orange each (Richard bought the oranges from a roadside stand) and an inch each of Snickers bar.

(I am actually writing this in first class in mid-Atlantic, courtesy of the *National Geographic*. I have just eaten two smoked salmon and salmon mousse canapés, one foie gras canapé, one dish of caviar with blinis, chopped yellow and white of egg, chopped parsley and chopped onion, *consommé julienne*, salad with radicchio, and brook trout stuffed with breadcrumbs, avocado and mushrooms. I am now on to the Coulommiers, Roquefort and grapes with vintage port. This annuls any self-improvement in Ranomafana.)

We hit their last movement to feed of the day. The subadult goes to his mum, and grooms his new sibling. I am much more conscious of identifying with these animals than I used to be, because I realize I don't just dote on the babies. I really feel for mothers hugging half-grown sons.

The social life of field scientists. Back at the camp, Patrick Daniels is making vegetarian stew. He cuts up one after another of luscious fresh greens, while people report their facts of the day.

Camp is mainly a great green tarp in ogival curves. A built stick table holds everything from Malagasy *sakay* (hot pickle) to genuine English Lea & Perrins sauce. This is expedition headquarters, and this is an expedition. OK, it's only half an hour's walk from a tarred road, but in just a year from the discovery of the red bamboo lemur Pat has pulled together a ganglion of scientists studying the structure of a forest ecosystem, deciphering facts never before known. And recognizing new species—new to science, new even as separate entities to the amazing guides.

Ken Creighton and Louise Emmons on sucker-footed bats and other small mammals, David Edelman (David Bamboo) on bamboo. Deborah Overdorff on red-bellied lemurs. David Myers (David Guitar) starting his thesis on red-fronted lemurs. Louise Emmons is radio-tracking two red forest rats, a male and a female. This mostly means sitting on a stump with her radio antenna getting three separate fixes on each rat for each observation time, and writing compass directions in her notebook. Rarely she sees one of them: diurnal, and handsome for a rat, a bit darker than the red forest floor. I saw one jump in two bounds over the path myself. They like bananas, mostly, which is endearing for what is, after all's said and done, a rat.[3]

Caroline Harcourt is just going out for the night. She hopes to radio-track avahi, the nocturnal woolly cousin of sifaka and indri. If she can find one and knows where it goes to sleep, then tomorrow we can dart it and fit on a radio collar.

Dick and Jen Byrne, visiting for the week from St. Andrews, Scotland, have been birdwatching. This is the forest where Dominique Halleux took magic photos of the endemic rainforest birds. It was Dominique and Lucienne Wilmé who originally trained the guides Loret and Émile Rajery. Dick Byrne was full of amazement at Rajery. He said Rajery must have seen and pointed out a dozen of the yellow-browed whatsit scuttling along the ground before Dick managed to spot it, and another dozen before Dick got his binoculars on one. Rajery and Loret aren't just

---

3. Sucker-footed bat: *Myzopoda aurita*. Red-fronted lemur: *Eulemur fulvus rufus* then, now a full species *Eulemur rufus*. Eastern red forest rat: *Nesomys rufus*. Eastern woolly lemurs have now been split into five species. The Ranomafana one is *Avahi peyrierasi*. *Atelornis crossleyi* is one of five species in the endemic bird family *Brachypteraciidae*.

sharp-eyes; they are scientists. Rajery did find Dick and Jen a good view of the pitta-like ground roller, which is a big bird but extremely secretive and hard to see, in spite of being red and blue and green all over like little Black Sambo.

This is the beauty of expeditions like this. Even if you have only a little scrap of data or experience to tell over the campfire, you can bask in other's appreciation. And there are always new eccentrics like Richard, brandishing his short-wave radio tuned to the BBC, and babbling of strange lands like Nicaragua and the Persian Gulf which must lie somewhere outside the forest.

At last the stew cooked. It seems that the double handful of green peas Patrick threw on top were in fact green Malagasy peppercorns. Even with rice it tasted like hot green coals. David Guitar played after supper, which helped.

So to bed in a borrowed tent—no mosquitoes, no leeches, no roots under the sleeping mats. Lovely.

Tuesday, August 4.

Pat stood on the palm-pole porch, face taut. 'I've just heard the worst news I've heard in a long time' says Pat.

Two foresters, *agents techniques des Eaux et Forêts,* walked the northern boundaries with the gang—it is their job if reserves are to be made, and they were the people who explained these strange strangers to the villagers. But *agent* Philippe left halfway through.

Philippe was in a very strange state. On the tour he was beset by bad dreams. At last he told Patrick he had too much money in his bank account, and was being audited. A neighbor in the same town had his wife put in jail and the stock of rosewood carvings they all make as a sideline confiscated by the police. Now they were auditing Philippe as well. Patrick made him repeat it all, three times, not believing he was really being told these things by a forest agent.

Please, pleaded Philippe, let me go home. By this time, though, the shortest way was to go on, and then take a five-hour shortcut back through the forest.

He never turned up.

And now Pat was holding a note, hand-written on squared school notebook paper, from the commissioner of police. It summoned those who had been on the trek with Philippe to appear and make statements about his disappearance. About his probable murder.

Meanwhile Caroline Harcourt had got up at 4 a.m. and tracked her avahi to its sleeping site. Only Patrick knew how to use the Cap-Chur gun. Please, please could he dart the animal for her before going to the police? Richard and Hubert went down to Ranomafana instead of Pat and Patrick to tell the commissioner that the team would come as soon as they could.

It was very odd to think of the dead man, and the multiple ramifying threats to the team and the reserve, while still in the Western intellectual cocoon of the field camp.

Patrick is tall and handsome in a blond, clean-cut ex-marine fashion. He was angry, and sweating, at the implications of the possible murder. Also, he said, he didn't like darting animals. He said this while kneeling in the sunlight with the Cap-Chur gun, mixing the doses of ketamine he calculated for a body weight of 1.2 kg, carefully pouring them into the syringe-tipped darts.

We set off: Caroline, Patrick, David Bamboo and me clumping after. 'They're really nearby,' Caroline assured us, which meant a kilometer or so up a particularly nasty clear-felled track, or clear-felled with stumps and vines, a swathe that the logging entrepreneur had cut when he thought he could bring in logging machines. Eventually found two avahi sleeping together as usual, only about 15 feet off the ground in the bushiest thickety bit of a small tree, wrapped round with vines and leafy branches. Well, you wouldn't expect small nocturnal animals to sleep where they could be seen.

Patrick circled gingerly round, carrying pack and gun (it's an airgun, and looks like a gun) to where he could see an opening. Very slow. The story is you must hit the animal's thigh. If there is any doubt, any risk of hitting vital organs, just don't shoot.

'Now' said Patrick levelly. There was a sharp bang in the silence, and David sprang toward the tree. 'I missed. Stop. don't scare it.'

David stopped.

The avahi had jerked up its feet with the bang but stayed put.

The pause seemed endless while Patrick reloaded, and took aim again, at the other animal, higher, to allow for that jerk of the feet.

Bang—the avahi leaped, leaped, leaped zigzag away from us. 'Keep it in sight' yelled Patrick. He and David and Caroline dashed after, me following, shedding packs and binoculars as they ran.

It had disappeared.

'I hit it square. We have to find it.'

We quartered the forest bushes and trees, and bushy vines.

'I've got it!' David shouted. It was hanging to a trunk, 10 or 15 feet up, helpless, but still hanging on. It probably wouldn't have fallen, just set its grip tight for an hour. Patrick grabbed the tree trunk and shinnied up. He took the animal gently in his arms, carried it to the trail, and laid it on a tarpaulin in the shade. Then he began to move with an extraordinary mixture of grimness and gentleness, and of haste and deliberation. He really did hate darting animals, hurting and scaring them, risking their death if he missed his shot. It was odd to see this big, muscled man tense with the care he had to take of the little furry creature. On the other hand, if it had to be done, then no scrap of data should be wasted or left out while the avahi lay there.

It was so small, with such human proportions as its long legs stretched out limp instead of folded for leaping. Fur grey washed with rufus, with a lovely white arched brow over its amber eyes like a worried child, and a little crown of bright rust red above each ear. It lay still—probably conscious, since that is what ketamine does, but unable to move a muscle in Patrick's big hands.

He checked its thigh—a clean wound, bleeding, but no more than a sore leg by this evening. He took its pulse with a digital rectal probe—steady at a hundred, not fluctuating with stress. He lifted it by a soft noose on the legs, attached to a spring balance: 890 grams, less than the weight he had dosed for. Small or young?

David Bamboo wrote the measures on a prepared data sheet on a clipboard as Patrick dictated. Anything to save time and stress. Total length. Tail length, which subtracted gives head and body. Chest girth. Leg, shank, foot length. Testicular diameter, left and right, which are never equal. Tooth wear. And then finger and foot prints, painted with graphite powder and pressed on a special scotch tape.

They put the animal in a gunny sack to take back to camp. Even though it was now starting to wiggle and kick, you have to keep such arboreal animals for several hours before you risk letting them go up a tree, leap and, if they are still groggy, fall.

Back at camp, the others were really worried. We'd been almost three hours with the distance, the chasing, the measurements. But Pat had her priorities: the avahi came first; the police could wait. She inspected the bright red collar, and held the soft grey white and rufus head in her hands and said, no, the collar was still too loose. It would come off and waste the animal's ordeal, or at worst catch on a branch. It should be just loose enough to slip a finger under, the exact measure being Pat Wright's little finger.

So, quickly, the collar was cut and fastened with two soft rivets that squash tight in their setting with a pair of pliers. Then the avahi was left to hide in quiet in its sack until time for release.

Patrick sat while Pat finished the job, shoveling in a delayed lunch of cold rice and beans, muttering, 'I hate darting animals. And I hate a lot worse what is coming next.'

Pat and Patrick went downhill with Rajery and Loret to find out how Philippe had died, and if, in any way, people thought them responsible. There seemed nothing that the rest of us could do, so we went back to our concerns, trying not to think. I went and watched red bamboo lemurs, now with the official name of *Hapalemur aureus*, given by all the people who saw it last year and the people who looked at its chromosomes to prove it was a new species.

Pat has the final joke. She has all three bamboo lemurs at Ranomafana! The big gray ones, *Hapalemur simus*, are sympatric with the red. It is the greater greys that make massive holes in the bamboo bases, with the spaghetti of splinters standing out at each end. The greater grays live in the small, distinct patch where that kind of damage is found. No one has seen the reds do more than bite off the tops to eat like corn on the cob. Sympatric species are real species. Hooray!

Back at camp. The avahi is due for release at dusk, and we're past that. Headlamps all round. The trail over all those miserable stumps looks even longer by moonlight.

Caroline has put a home-made red plastic cap on her headlamp. Richard sits patiently behind her with private thoughts about how glad he is not to be a zoologist. The avahi sits in front, only about 3 feet off the ground. You can see its sore leg, even by red light, but being nocturnal it doesn't see red light itself.

The avahi's female has disappeared. He is all alone, unless you count five great humans.

After half an hour he slides down the tree and sits immobile on the ground.

After an hour Richard reminds us that dinner is supposed to be at 8.

Caroline, who rose at 4 a.m. and has been with this animal ever since, says she's not leaving till she knows it's all right. At this point it looks like prey for any rambling little carnivore, let alone the dreaded fosa.

Richard says, what about a deal? If she's not back by 9, we'll return with supper. 'Fine' agrees Caroline, without taking her eyes from the avahi.

Everyone is there for dinner except Pat and Patrick, who'd said they'd stay over in town, either in the hotel or in jail. We eat David Bamboo's curry, then very reluctantly start back with bread, bananas and good Swiss chocolate toward Caroline. We have all the glow of virtue and none of the work. She's coming back whistling, since the avahi sat up and ate. She has certainly earned her chocolate!

In the morning she tracked his little beeping collar to his sleep-tree, and found him cuddled up with his female.

Aug. 5. Murder.

It seems that the police interviews were straightforward. The police just wanted a statement from Pat and Patrick. They tone down all mention of Philippe's nocturnal hallucinations, just saying he had both family and financial worries. We talk of the likelihood of suicide.

Now Pat and Patricia and the guides will go to pay respects to Philippe's family. The guides change to funeral clothes: Loret in a red Duke T-shirt with sifaka face, Rajery in a pale turquoise version, with matching turquoise corduroy cap.

After too long and too good a lunch—all praise to the French Mission to Civilize, in its culinary sphere—Richard and Hubert and I start driving downhill toward the old coffee station of Kianjavato.

Savoka is supposed to mean the second growth which sprouts where primary forest has been cut. Below Ranomafana it means desolation—bare hillsides with grass, bare hillsides with low weedy bush and an imbecile's topknot of scraggly trees, bare hillsides regrowing bamboo in graceful hoops and arches. This is not a poor area, for Madagascar. People sell bananas and pineapples that are trucked up the road to Tana, and there is a fairly continuous sprinkling of roadside villages. There is something here to develop.

On down, to the obligatory camera stop at the top of the second escarpment step. A wide panorama of a few houses, a few fields, bamboo stands, and empty scrub on empty hills. Still further down there are no more villages, just travellers' palms and bunch grass on sterile soil. It once held forest, and there is still enough rain here for rainforest to grow. This is man-made sterility, inexorably creeping uphill toward the glorious remaining forests of Ranomafana.

As we buy gas at the woodshop, Hubert hears from the shopkeeper that Philippe's death was not suicide; it was indeed murder. The last village he passed is under deep suspicion: to them, he was rich, worth robbing, or else they are so isolated they were frightened of the stranger. It is said the men chopped him in pieces and hid the pieces in the forest. When the police came to ask questions a village child was so terrified he admitted seeing it all.

Will this mean anything worse for the researchers of Ranomafana, so careful of the little avahi, but floundering in the mysteries of the human world?

FOURTEEN

The village of the fig tree

> We are Tanala, which means the people of the forest.
> We are like the lemurs: the forest is our home.
>
> *Lord Ravanomasina, July 1, 1993*

July 1, 1993, Ambodiviavy village near Ranomafana.

I sit with Bob Brandstetter, head of the USAID Evaluation Mission of Ranomafana National Park, and Jean-Marc Andriamanantena, of USAID, in the house of the Mpanjaka Zafy. *Mpanjaka* means king: Zafy is traditional king of the village of Ambodiviavy, 'the village of the fig tree'. Rain pelts down outside and runs off the thatch, as usual. Three out of the four village nobles are seated on a rolled-up straw mat, which raises them maybe three inches above the rest of the crowd jammed into the mud-walled room, but we visitors have real chairs. The three men are studies in rosewood-colored skin, high cheekbones, short-cropped graying hair. The Mpanjaka retains only two yellow teeth. Next to him his representative, Zanaka, is in strong-built middle age. He is the most outspoken: his role as representative is to speak. The third, the oldest, Ravanomasina, says only the phrase I quoted above as a coda to close our session. He is actually the ancient royalty of the region, whose title dates from before the Islamicized influx of four centuries back. This silent, ancient king is called 'Andriana,' 'Lord,' the newer one 'Mpanjaka' or 'King'—a dual royal lineage, carried on through generations of men who sit huddled against each other on the same mat.

These three nobles are wrapped in *lambas*, checkered cotton cloths, none too clean, that drape with fluid majesty. Now as they sit they are converted to carved square shapes like that classic statue of the scribe from the earliest Egyptian dynasty, topped by the rosewood heads. A fourth enters later, a huge man with long black limbs, dressed in traditional woven raffia shirt, woven palm-leaf rain vest, and square palm-leaf cap. He is Lesonina, the scribe of the village, who can read and write. He keeps books of village work hours in copperplate hand—they have a detailed, written system of work-sharing and benefit-sharing. Finally, further back among the crowd, sits a visiting Mpanjaka called Raymond from another village.

Why is there a park? After the usual protracted courtesies, Bob Brandstetter begins with the crucial question: 'What does the new National Park mean to you?'

Zanaka fires back: 'Nothing. We don't know what it means.' Very well, if he is stonewalling, we will circle round to other subjects, and return until we understand him.

Ambodiaviavy is a cluster of mud huts, only twenty minutes' walk from the potholed Route Nationale. It looks like a timelessly isolated Tanala village, but the old king did his military service in the capital, another man ran a small business in the town of Fianarantsoa, and all have relatives on the plateau above. Several villagers have mixed marriages, not with other Tanala but with Betsileo from the plateau, where people do only settled paddy farming, not slash-and-burn like the forest-edge Tanala.

We ask who has left the village and seen what life is like in other regions where people have abused the ecology and lost the fertility of trees and soil. They indicate Raymond. He went on a study tour organized by the Ranomafana Park Project. He traveled all the way down to the coast, over bare, denuded grasslands, to see what it means to live without forest. Raymond is now the authority. Bob questions him. Raymond says it is miserable to live without trees. Trees are the savor of the earth—trees give the earth its taste, and fertility for cultivated crops. Since his return the villages have started *reboisement*, tree planting. What do they plant? Eucalyptus and fruit trees... We didn't press them, and it is not too clear that this is actually happening. But Raymond wants to start a tree nursery. Why? Not too clear... But it is the right answer. Still, Raymond wonders. They have been told that with no trees there is no rain. When they look on the plateau, there is still rain, and there are still crops.

I explain that there is less rain, not no rain. This is a terrible problem where I work, in the South, where your tongue goes dry with thirst. Here at Ambodiviavy there will still be rain. (Indeed, I can scarcely remember what it was like not to have rain in Ranomafana.) The real problem here is erosion and silting of rice fields. Uphill on the plateau, the problem is wood itself. They have to use only mud and brick for houses, have to buy a few timbers for the roof. The problems that come from deforestation differ in each region, but there are always problems. Gabriel, the Park Project education officer, volunteers that it costs 50 francs for three sticks of firewood in Tana now. The Tanala look a bit skeptical... I don't think they want to imagine a place like that.

Zanaka, the king's representative, turns out to be not only outspoken but willing to contradict. Raymond may say trees are the savor of the earth, but he, like Raymond, has seen places where people live without trees. If they live without trees on the plateau and the coast, the Tanala could live without trees right here!

Bob tries a variant of his first question: why has the Park been declared? Zanaka answers that it is a park for *vazaha*—foreigners. So be it. There is land outside the park which they have been told is still accessible. However, to do slash-and-burn (*tavy*) even on this land they have to travel all the way to Ifanadiana for permits from the Eaux et Forêts forester—hours on foot, expensive by *taxi-brousse*. When you look across the valley and up the other side, perhaps forty minutes' walk, the top third of the mountain is beautiful high forest, shaved off sheer, like commercial clear-felling at the sharp edge. There is second-growth below, but only 4 meters high at most—too short for sustainable fallowing, though we did not ask. No way should they be cutting the primary forest for further *tavy*—but they quite reasonably say that they are the people of the forest. They say they know how to manage forest, what trees to leave, how to burn, how to rotate the *tavy* when it is no longer fertile. One has to sympathize that they are now treated like children who cannot know as much about their own forest as a bunch of foreigners. 'Foreigners,' incidentally, includes the Water and Forest Department, who are mostly plateau people.

Biodiversity. I ask about lemurs, as the one group where my knowledge of biodiversity might exceed theirs. They name four of the twelve kinds hereabouts. The simpona, the chocolate and white ballet-leaper, and its nocturnal cousin the avahi, do no harm. The varika mena eats their bananas (they are classic opportunistic raiders). The varibolo, or bamboo lemur, eats the new shoots of their mountain rice, and is a real pest.[1] I follow up enough to be sure this is the common gray bamboo lemur, not its rare congener, the greater bamboo lemur, or the ultra-rare golden bamboo

1. Simpona, Milne-Edwards' sifaka, *Propithecus edwardsii*. Avahi, eastern woolly lemur, *Avahi laniger*. Varika mena, *Eulemur rufus*. Varibolo, gray bamboo lemur, *Hapalemur griseus*.

lemur. Only a dozen or so individual golden bamboo lemurs are known to exist—all in this national park.

Land tenure. I ask how far the villagers have to walk now to get forest products. It is only half an hour for construction wood, but four hours to gather the medicinal herbs that grow in the high forest. Bob asks if they ever go into the Park. They say no. He says 'Just between us, you still do'… to which they give the only possible answer: 'No, we are too afraid of the law.'

The Park is forbidden, although the land of their ancestors extends into the Park. Furthermore, there has been a cadastral (surveying) team here two months ago, which tried to establish boundaries of personal property. This is DDRA, Direction de Reform Agraire, fruit of François Falloux's World Bank dream of 1987. They are apparently only surveying rice fields whose ownership is already clear, and not touching the *tavy* fields, whose stabilization was Falloux's original goal. The villagers confirm this, which increases their fear that the invading government will leave them with only bottomland, which is not enough to live on. Furthermore personal property is against all tradition: the village lands should be registered in the name of the king. The notables murmur assent, especially Mr. Outspoken Zanaka. Jean-Marc of USAID is indignant. The ethnographer in him rises up at this breach of tradition by the directorate of land reform.

It wasn't till later that I thought to ask just whose interests are represented, and found out that Mr. Outspoken is not only the representative but the inheritor from the king, and the notables are essentially all the king's men. Most of the village are related, but it is a question in my mind whether this means distribution to all. It seems there is a second group, called *vahiny*, or visitors, who have only been around for a few generations… Heaven help those who try to sort out land tenure, including the peasants themselves.

Women of Ambodiaviavy. The Women's Association is represented by one older woman of intricate oiled knotted braids (Rahasoa, whom anthropologist Paul Hanson jokes really runs the village) and one young mother, the lord's daughter, with a nursing baby who has a round butter-soft face and huge baby seal eyes and a little string knotted through her ear where she will wear an earring. I spend some time 'alone' with these two women plus interpreters. I try for questions but they are far more outspoken than the men. We need medicines. Could I treat elderly Rahasoa? Village health aids have no legal right to buy drugs for dispensing, except chloroquine, by order of the Ministry of Health. The grounds are that they would not know how to prescribe them. There is a Unicef pharmacy in Ranomafana town which sells drugs at 20 francs a pill. Individuals can only buy the drugs with a doctor's prescription, and then only one course at a time. Meanwhile the women point out that there are people with dysentery in the village. They know all about common diarrhea and oral rehydration therapy, of course. This season of cold, continuous rain brings true bloody dysentery. They are well aware that for dysentery they need tetracycline. Could I please *not* ask them all the questions all the other visitors ask, which takes a lot of their time, unless I can do something concrete about keeping my promises. In any case Rahasoa announces firmly that she has had no children of her own, only adopted children, so she can't enlighten me about family planning. It makes me think of a minister receiving his fifteenth World Bank Mission of the month.

Farewells. As we leave, the most ancient lord concludes with that phrase: 'We are Tanala, the people of the forest. We are like the lemurs: the forest is our home.' Our little group treks uphill to the beautiful two-roomed house the village built for resident grad student ethnographer Paul Hanson. We admire through the silvery rain-streams Paul's view of the square rice fields of the valley, and the high forest towering above tufted *tavy* cuttings. Rahasoa, the older woman, walks uphill with a pot of steaming sweet floury manioc. It tastes like roasted chestnuts. Only chestnuts are never a sign of such warm and welcome hospitality.

Afterthought on the mud trail back. These are not such poor peasants as Malagasy go. Several girls had gold earrings, and the young head of the Women's Association, the Andriana's daughter with the beautiful baby, sported two gold teeth. They still have a little fertility in their soil from the forest, and cattle pastured in hiding in the forest. Simultaneously, almost, Bob Brandstetter remarked that he has visited countries from Mali to the Sudan, and never seen poverty like Madagascar.

July fourth? or fifth or sixth?... Time blurs in Ranomafana. Zanaka has requested a formal and private interview with Bob and me. There were so many project people in the village that he could not talk freely—in spite of my dubbing him Mr. Outspoken.

We meet privately, then, with Paul Hanson to interpret, in the Hotel n'Kanaka. This is a cubical shed of corrugated iron with open spaces for door and windows, large enough for five wooden tables. Zanaka has a fine blue rain slicker that may derive from the Park Project, and whose hem is just low enough to make him seem otherwise unclothed.

He comes straight to the point. The Park is treating villagers wrong. They have their own structures. The Park should give them money for fertilizer and seeds, not loans but gifts. They would buy seeds of pineapples and potatoes, and little fruit trees, for the fallow rice fields and the *tavy* fields. Bob asks, 'What would they do with $100? And Zanaka shoots right back, 'Buy pigs!'

However, any gifts of money should go to the Mpanjaka for the whole village community, not to individuals. You should always give village money to the king, but it should be done formally, in front of everyone, with the sum stated loudly and clearly and the bills shown in front of the general village council. Then it is a village pride, and the king shares properly according to work done or village priority. It would be *baraka*—shameful—in such circumstances, for one individual to keep it all.

Zanaka has come to make a formal request, as well, for medicines. Nothing has yet been accomplished. He did not wish to state this in front of everyone, as it would be *baraka* to request that which might not be given, and which it would shame us not to give.

Bob Brandstetter returns to that strange answer he gave before to the one central question, that he does not know what the Park means. Zanaka insists it is true. The Park team came and had a meeting, but they gave no real explanation of what the Park means. Furthermore, he has relatives in most of the villages on the eastern side of the Park, and he can testify that none of them knows what the Park means.

What he does know is history. He tells us the story of their ancestors' land. Ambodiviavy is the fourth village. The first, Ambatofotsy, was beside the road. The land was delimited and claimed by a French colonialist, so they were moved. The second, the same. The third, on a hillcrest within sight of here, where the school stands now, was burnt by the French in 1947, and many were killed in a *razzia*. (*Razzia* is an Algerian word for a raid into enemy territory—the French borrowed it for what Americans later called 'search and destroy.') The survivors hid in a cavern in the forest, a sacred site now out of bounds inside the Park. They stayed for four months, sneaking back at night to gather rice from their village fields. Then, in 1948, they came and settled here. Ambodiaviavy means 'the village of the fig tree'—but there is no fig tree. That grew in the site they lost, the third village, which was burned.

Would it take only an hour or a year with this forthright man to find out what he thinks the Park means? There is terror of the land surveying and the boundary posts that mark out only permanent rice fields, not *tavy*, and even those all wrong as belonging to individuals, not clans. There is the Park that promises help but does not bring medicines or seeds or fertilizer, which he knows they need. There is the history of colonization. Maybe the reason he does not think we have said what the Park means is that we talk about trees and lemurs and a future fifteen or twenty years away, when their *tavy* will have destroyed the whole forest if they do not stop cutting. Perhaps Zanaka is waiting for us to admit the one central fact the villagers do know... that once again they have lost their land.

Zanaka walks away down the road: long brown bare legs, shiny wet blue jacket, rain pearls on his grizzled hair, back toward the village of the fig tree.[2]

2. For another view of an ICDP from the bottom up, and the subsequent confusions up to the present day, this time in Mananara Nord, where Eleanor Sterling and I saw the aye-aye, see G.M. Sodikoff, *Forest and Labor in Madagascar: From Colonial Concession to Global Biosphere* (2012).

FIFTEEN

Development meltdown

Wednesday, July 7, 1993, Ranomafana.

What is so hard to explain about the drama at Ranomafana is that no one actually cares whether expatriates have affairs. In Madagascar, as elsewhere, injured parties may storm or fume or poison each other. A mere affair, however, would normally raise no ripples in the community at large, let alone lead to expulsion by USAID with orders not to return to Madagascar. The one point that is totally irrelevant is whether Dai slept with Dennis.

I suspect it has something to do with the rain. Clothes and papers mildew, but tempers rot.

The real substance is probably about tribal jealousies, job jealousies, the administration of Madagascar's most ambitious project to combine environment and development, with the tensions brought by our evaluation team come to judge whether $3 million of USAID money has been well allotted and well spent.

Six years ago I first came to Ranomafana on the day the diadema fell down. I remember the golden sunlight, and trails stair-stepping up the rainforest escarpment and the winsome face of the unknown bamboo lemur. I remember also the early rivalry between Patricia Wright, who first found the golden bamboo lemur, and Bernhard Meier, who with Roland Albignac named it as a new species.

Now I am back, to my own surprise, evaluating the grandiose regional project which Pat Wright has created. In May 1991, Madagascar officially dedicated 416 square kilometers of the region's forest as a National Park. The Park is backed by a USAID grant to support conservation and to provide villagers with an alternative to the slash-and-burn farming which sends their biodiversity up in flames. The Park, the Project, and another $2 million of foundation support are almost wholly due to Patricia Wright. She fell in love with the forest. She has attracted specialists to study its sucker-footed bats and endemic silver-sided fish. She has launched television teams up the steep trails to film the panorama of treetops that falls away to barren plains below where the earth is already exhausted. She has also fallen in love with the people. She charms the old *Mpanjaka*, the traditional village kings. She hires ambitious young men and turns them into research guides who are among the most knowledgeable biologists of Madagascar. If there is a Ranomafana National Park, and a USAID-sponsored Park Project, thank Patricia Wright.

I've known Pat a long time; her swinging black hair, her infectious grin full of crooked teeth, and some of her lovers. I am still annoyed with my husband for saying Pat gave the best lecture on integrating conservation and development that he ever heard—even better than mine. If I had a spare million dollars, Pat might get it.

I was distinctly flattered when Pat asked if I could come write the biodiversity research section of the Evaluation Mission for the RNP Project. I was also pleased that C.J. Rushin-Bell of USAID in Antananarivo enthusiastically supported me as part of the team. I can write about biodiversity, of course, but have no experience of grant administration. C.J., as administrator of USAID'S largest biodiversity program, has to be tough-minded about appointments. For relaxation she goes out on weekends with the duty station marines and teaches them fancy rock climbing on Madagascar's granite cliffs.

There seemed to be overtones in the air when I took my place in the chilly hall of the Hotel des Thermes de Ranomafana. We listened to presentations about fishponds and health care, and an exhortation by Dennis del Castillo, an expert from North Carolina State University about the poverty of the soil and the need for composting.

Dennis is short, dark, intense and Peruvian. His French and English are both easier to understand if you already speak Spanish. He seemed on the defensive, a missionary among heathens. He won me at first by showing that cartoon I've often redrawn from the *Midi Madagascar* newspaper: a ring-tailed lemur, its tail stuffed with thousand-dollar bills, promenading past a couple of starving Malagasy and remarking 'Tough luck! You're not an endangered species.' Dennis said, as if launching a challenge, 'This may be a Park Project, but the Park will never survive without the agreement of the people.'

He showed figures of soil minerals. These ancient laterite soils are among the worst in the world. The phosphorous content is vanishingly small, and nitrogen not much better. The only hope in a place too poor to buy fertilizer is total recycling.

The local Malagasy audience, all of them project personnel, shifted uncomfortably, knowing what was ahead.

Total recycling, Dennis went on. All organic waste, including human waste. People should defecate under their coffee bushes or orange trees, and scratch a little dirt over the feces, like cat latrines. All very well for the health team to promote deep pit latrines, to avoid diseases. Most people don't use them anyhow—so why not deliberately enrich the crops?

'If I told what you said to the ministers in Antananarivo, you'd be thrown out of the country!' Jean-Marc Andriamanantena was white with fury. Jean-Marc trained as an anthropologist and now is on his fifth evaluation mission for USAID—the AID member of our evaluation team. All his inter-cultural sophistication evaporated when he found himself opposite Dennis at lunch. 'Feces under a fruit tree, when you plan to eat the fruit! Or to poison your neighbor with it! I counsel you very strongly to give up this idea, and to respect the people you are trying to work with!' (A furious Malagasy of Jean-Marc's Indonesian stock does not literally go white, but ominous steel blue about the lips.)

Dennis flung out his arms in Latin bravado. 'I promise you, I do respect the farmers. I know what they face. I grew up on a hill farm. I had my first pair of shoes when I was 15! I know as well as anyone what is needed here. And I tell you these farms grow nothing—*nada*—unless you enrich the soil! We tell them not to do *tavy* on the hill slopes, but we must give them some real alternative!'

They glared at each other: Jean-Marc, whose name itself indicates aristocratic lineage, and this continent-hopping interloper who wanted the Malagasy to foul their food.

Bob Brandstetter, head of the evaluation team, whispered to me, 'We are all going to dinner with Dennis and Joe and Dai tonight to talk about the troubles.'

'Troubles?' It was just my first day.

'Oh didn't you know there was trouble? Welcome to the team!'

'What's the trouble?'

'Something between Duke and Stony Brook, who administer the AID grant overall but concentrate on research, and North Carolina State, which has subcontracted all the environmental development side. Dennis and Joe and Dai are all from NC State. Dai isn't official of course—she is a graduate student doing ethnography, on her own grant.'

'Should I have known?'

'I heard about trouble already, back in Washington. Anyway, we'll all find out tonight.'

After lunch we all filed back to hear about biodiversity—the section of the project administered by Patricia Wright herself. In her usual spellbinding style she told us how the project had grown from her first discovery of the golden bamboo lemur in 1986 through the tented camp of later years, to the present solid log cabin where twenty researchers may join at once—half foreign, half Malagasy—to eat their communal rice and beans. From there they fan out through the escarpment forest to mark mouse lemurs, monitor plant plots, bound along beneath the leaping sifaka, discover and name species after unknown species. Twenty-four Malagasy students of science or economics have already earned their Master's from work in Ranomafana. Pat modestly does not mention her own research, which is most economically significant of all. She has shown how lemurs spread the seeds of their own food trees, the high, old, valuable trees of the forest. Lemurs are in effect the nursery gardeners of Madagascar's biodiversity. This story would have tied into Dennis's proposal—even a little plop of lemur fertilizer is enough to help a seedling grow.[1]

The twelve research guides, an elite corps whom Pat has trained, sit in a row in their green coveralls and wellington boots. Their leaders are still Émile Rajeriarson and Loret Rasabo, whom I met long ago on the day the diadema fell down. Loret is still keenest on birds, but Émile has turned from lemurs to a new fascination with the role of forest ants

1. J. Dew and P.C. Wright, 'Frugivory and seed dispersal by four species of primates in Madagascar's eastern rainforest' (1998).

and wants to publish articles himself. Disappointingly when it is their turn to speak, they ask me about off-site per diems and study tours. (Later on, when we talked more, it seems this is really about career structure and recognition. Having made a transition from semi-literate hunters to research personnel, they would like a few of the accompanying perks.)

On the whole the Duke–Stony Brook, or biodiversity, side of the project, as championed by Pat, sounded pretty good. I was staying with Pat in a modern but smoke-filled house five minutes' walk from the hotel. We splashed to the house in the cold rain—jacket and wellington boots already picking up the dank smell that persisted for the next two cold, rainy weeks.

A car came to fetch me for dinner—a few minutes' drive in the dark in the rain was enough to lose me completely. I came into a wooden, straw-mat-lined room in what I knew was Dennis's house. It was a square box of electric light suspended on piles like a stage set in the night. Black rain surrounded it on all sides; oily black rivulets oozed through the black mud below.

Our evaluation team of four sat to eat. (Jean-Marc, fortunately, did not come.) Bob, the team leader, had evaluated other big projects in Madagascar. The surge of interest which came with the Environmental Action Plan has blossomed into the so-called 'Integrated Conservation and Development Projects,' ICDPs, all over Madagascar. However, many so far seemed to be disasters. There was a chief of party in the Montagne d'Ambre who wouldn't even meet the evaluation team that came to judge the project's performance. Masoala never got off the ground: the principal conservation agency has now changed abruptly. Andohahela flunked too; the Liz Claiborne Art Ortenberg Foundation withdrew their grant after reading Bob's verdict.

We are supposed to approve of Ranomafana. This is the showpiece where every sector claims progress. Ranomafana is the place where biodiversity is said to be holding hands with human development.

The other two members of our team are exceedingly distinguished. Donald Stone of the Organization for Tropical Studies heads up the fifty-university consortium which runs La Selva, the long-standing research station in Costa Rica. Hugh Popenoe has a down-home Southern accent and manner, an encyclopedic knowledge of tropical agriculture, and runs fifteen-odd AID grants like this one from his office in Gainesville. Neither Donald nor Hugh speaks French—but they have picked up all the essentials, including Jean-Marc's view of Dennis's presentation, which happened to be said in English.

Dennis is a welcoming host. There are a dozen people in the room besides our team, including Dai and Joe Peters, a dour-looking student called Dan Turk and several Malagasy who work in agriculture. The common link is that all are employed by North Carolina State University, subcontractor for conservation and development. It seems a jolly dinner, mixing Malagasy and Mediterranean dishes. Dai has even cooked Chinese fried rice. In spite of Bob's warning, I still thought they were being just hospitable.

I happened to sit by Dai. She is tall for a Han Chinese, with a loose-limbed stride that implies that even if she started from south China she spent formative years in American blue jeans. Her face is very flat, with high cheekbones and thin nose—a face that is not pretty, only beautiful. Her hair waves outward like a cape over her shoulders or a black nun's cowl. Dai said that if we had time when the talking was done, she would love to show us her videotape. She said she was the only one who really knew the villagers. She spent three months in each of four villages—two uphill from the reserve and two down. Some people claimed she did not know villages from the inside because she was only a sociologist, not a committed anthropologist—but she did, she did.

Mellow from fried rice and moussaka we turned to Joe Peters, Dai's husband. He rose clutching a sheaf of papers in his hand. 'You all know why we asked you,' he began. (Not me! a little voice squeaked in the back of my head.) 'You know there has been friction between Duke and NC State. I think I had best give you documentation.'

He then read, with dates, from several letters, which chronicled the breakdown in communication between the two institutions. I understood nothing—it was all too oblique—except the denouement. A month earlier Thurman Groves, chief administrator for NC State, wrote USAID to say that NC State would not renew its contract if subordinate to Duke and Stony Brook.

Joe's voice began to tremble with passion, and his thin face to grow whiter. 'You have heard our public presentations today and yesterday. You have heard what Dennis has done with agriculture—fishponds and low-cost homemade weeders and his model farmer who has increased her yield tenfold. You have heard how I have dealt with Park administration and laid the basis for profitable tourism. You have heard how our graduate students explored the aspects of conservation that actually touch the farmers, like Dan Turk here with his forest tree plantations, or indeed my wife Dai who is studying human relations with the forest.

'I know I sound emotional. I am an emotional man. But I ask you straight—can the Park survive if the people do not come first? Can this project go on having hit-or-miss administrators who mainly care about research in the forest, instead of people? I think you must agree there is only one solution. NC State should be lead institution in Phase II of the AID grant. NC State should lead because we are good administrators, and because in this project, people have to come first!'

What could we say? I reached Ranomafana today, the others two days back. Bob said we'd look into the situation—that's our job. When the talk frittered out I asked brightly if we might watch Dai's video, for dessert, just to bring us back to the peasants themselves. We clustered around the screen like US householders, not a cast of players on a lighted stage afloat on the muddy waters of Madagascar's rainforest.

Dai's plot was simple. Uphill, above the forest, live Betsileo tribesmen. The Betsileo do not cut forest; they irrigate flat paddy fields, and fish for endemic crayfish that they sell. They are organized, thrifty, hard-working. She showed them plowing the heavy laterite with zebu-drawn plows, and pumping a two-tree-trunk bellows to forge iron.

Downhill, Dai showed Tanala tribesmen, the 'People of the Forest.' They cut slash-and-burn *tavy* fields higher and higher up the slopes. The lower slopes are incredibly steep but they aren't the almost sheer cliff that hems off the Betsileo from the forest at the top.

Dai showed us a Tanala man singing as he hacked down a tree for *tavy*, and Tanala families singing as they mourned a young man killed in traditional bull wrestling. She showed the relatives reeling back and forth down the road with the wrapped but stinking corpse. 'They play with it all the way to the grave,' she explained. In the month before Independence Day the remote Tanala village where she stayed danced every night. 'I don't know why Independence Day matters so much to people out here!' said Dai. (I do. After the revolt of 1947 the French burnt villages in this region, killed men, women and children and resettled survivors in hamlets where they could be controlled—an unnoticed precursor of our own Vietnam policies. Independence means much to the Tanala.)

'Anyway, she went on, they are happy people. They dance half the night, and then go sleep with whomever they please. They still have forest, so their soil is rich. They just set up *tavy* fields. The real hard work is cultivating rice on their few bottom lands and the Tanala are rich enough to hire Betsileo to come down and do the hard work for them.'

The picture cut to a woman beside a pool. She was urinating, standing up with her skirts held out around her pregnant belly while her toddler sat wide-eyed in the sling on her back. The woman and her dress and the pool were all the color of mud; the stream of urine glinted silver between her legs.

'They are happy, and they just don't care about things,' continued Dai. 'This pool is the village drinking water. They brush their teeth with that sand at the bank where she is peeing.'

Then we cut back to the plateau Betsileo with their frugal travail. Dai concluded that it was somehow paradoxical that the Park's whole goal is to replace the cheerful Tanala ways, based on unlimited *tavy*, with the settled Betsileo life. In the *soudure*, the linking period when one rice harvest has run out and the next has not yet grown, Betsileo lose weight and many children are critically malnourished. The Betsileo deserve better, considering how hard they work and how quickly they learn progress, but they are chained to their unyielding fields.

Out into the wet night. 'What did you think?' Bob asked.

'Amazing they let her take those pictures.'

'That film was the trouble I heard about in Washington.' Bob said. 'Apparently Dai showed it to some ministers. They have banned her taking it to the States. That is hardly the view they want to convey of Madagascar!'

Thursday.

I woke up at 3 a.m. Rain pattered on the tin roof and dribbled from the eaves. My bed was comfortably hard, but the blanket smelled damp. I spread my gray and white ski sweater over my feet, and replayed the film in my mind.

Village women do not bathe with village men, at least where I work in the south. They go down separately to the pool. Women do not urinate in public—elaborate codes rule village life. Only codes of manners can produce any privacy among the huts of sticks or mud.[2] Women don't sleep with any partner they happen to find. Their sex lives are hardly like the home life of our own dear Queen Victoria—but might not shock the present generation of royalty. And those 'happy, carefree' Tanala have an infant mortality rate as high as the nose-to-the-grindstone Betsileo. Dai's video is the start of a general description, though the reality is of course more complicated.

Dai had looked at human beings with the unblinking eye I try to use on lemurs. And then she showed film of a woman urinating to ministers of the government—most of whom would be men. Indeed, Merina men with their own strict code about toilets.

I thought of how to find Dai in the morning to point out that she was in deep shit—not a phrase I usually use.

Too late. In the morning C.J. Rushin-Bell sat by the breakfast table while our hosts fed sticks into the smoky fire.

'C.J.! Great to see you! What brings you to sunny Ranomafana?'

C.J. is one of the trimmest looking women I know—short but sculptured hair, quick step, quick smile. No smile today. She looked grim as one of her own rock faces.

2. I have wondered since if that woman was so pregnant that perhaps she could not help peeing by the pool and in front of the camera.

'I've come to drop the axe on the people from NC State.'

USAID had received letters of complaint from a Catholic NGO, and from the president of the province or *faritany*. They claimed that Dai and Dennis were carrying on an affair with public behavior that shocked the constituency. Dai and Dennis, and by association Joe, are unacceptable Americans.

If confirmation were needed, it just so happened that the head of USAID came to visit Ranomafana a few weeks before. He spotted Dai riding pillion behind Dennis on his motorbike, wearing shorts 'cut up to the armpits.' The head of AID tried to avert his children's eyes unsuccessfully. I could well imagine Dai with long bare golden legs and the cape of black hair streaming out behind. In this country women wear skirts. Miniskirts do exist in Madagascar, though not in Ranomafana. I've seen miniskirts on high-priced prostitutes that hang out at the capital's best hotel.

'It doesn't matter what they do in private,' concluded C.J., 'but if they are unacceptable in public, they are out. It is simpler than that, though. Their chief administrator has written to say NC State does not wish to take part in the Project's Phase II if it is led by Duke and Stony Brook. We just accept NC State's resignation. There is no reason to target individuals. They are young, with careers ahead.'

I had a vision of Joe last night triumphantly reading his copy of the ultimatum from his chief administrator. Joe had thought it would win his faction control of the project for the next five years. I seemed to see the paper turning to ashes in his hands, like a contract with the devil.

C.J. set her jaw, pulled on her rain jacket, and went off to tell Joe and Dennis that their contracts would end in eight weeks. They'd counted at least on a normal six-month extension, until December, if not the five-year Phase II. And then it turned out that by a confusion of clauses the NC

State contract stops on August 15, not August 31. 'Oh good!' C.J. told them. 'You can leave even sooner!'

C.J. herself is off in another ten days to take over USAID's environmental concerns in all Eastern Europe, including the blighted zones of ex-Soviet states like Kazakhstan. I guess USAID thinks she's tough. I wish I were tough. Bob and I squelched through mud today, only 20 minutes off road, to an interview with village elders in Ambodiviavy, the village of the fig tree. In this village there are Betsileo married to Tanala. People are well aware of all the possible kinds of agriculture uphill or down, and the women are so bored by well-meaning inquiries that they warn me not to waste their time with empty promises. They are obviously sick of foreigners' fascination with birth control. They don't fit into anyone's generalizations, including Dai's.

It's better to be wet in a village than a town, at any rate—the streaming hut eaves overhang so you can shelter, and the old woman who was so curt about birth control brought us hot sweet manioc, like roasted chestnuts.

Friday.

'Rumors! It's all a can of damned rumors!' Roy Hagen was irate, a chronic condition as opposed to Jean-Marc's earlier fury. Biology textbooks tell you the opposite action of sympathetic and parasympathetic nervous systems. One produces cold, blue-white anger, the other flushed apoplexy. Roy's was the pink sort. He is a big, blond man, with Viking bristle to eyebrows and mustache. Roy came down from Antananarivo to represent ANGAP, the Malagasy parastatal organization set up under the Environmental Action Plan to run the National Parks and Reserves at the insistence of the World Bank.

Roy declared: 'I've been staying with Joe and Dai and Dennis for days now. I'm as sure as anyone can be there is nothing going on between Dai and Dennis. That's just a pack of lies. There's all sorts of other stories, too. For instance, they are said to be hunting for gold—because they take soil samples! In fact, so many rumors are bubbling up about those three that I presume someone is spreading rumors. There is no other way to discredit the good work the NC State people are doing.'

'Roy, who would bother? I mean, this country is always full of rumor—but you make it sound like a conspiracy.'

'Look at it hard. If NC State doesn't deliver on development projects here, who really loses?'

'The peasants?'

'Oh sure, they lose. But who loses at high level?'

'I'm sorry. I'm no good at political games. You tell me who loses.'

'I'll tell you. ANGAP, That's who. USAID is pulling the rug out from under us all the time, disregarding what we say. The thing is, we stress that the crucial part of all these integrated projects is the people side. The NC State people are doing all the organizational and development work here. ANGAP knows that. We would back NC State to run the whole project efficiently. A biodiversity research cabin in the woods isn't going to save this forest.'

'Hey, wait a minute. This project wouldn't exist without Pat, and without the research. The explicit goal of the project is to save biodiversity. For that matter, if Pat had found the golden bamboo lemur in some other forest, the whole project would be somewhere else, not Ranomafana... Anyway, I still don't get it. Are you saying that someone is starting wild rumors about the NC State people just to discredit ANGAP? It doesn't hold water.'

'Remember how many people in the government do not want ANGAP to become a real National Parks service. And remember how vulnerable we are. Take your evaluation team. AID sprung that on us with three weeks' warning, all complete, as if we weren't even worth consulting. You're lucky we accepted the members at all. Two of the team speak no French and have never been in Madagascar.' (Roy did not add, out loud, 'and one is known as a friend of Pat's—hardly a fair evaluator.')

'But even if someone is trying to get at ANGAP, it seems an awful long way around to start rumors in Ranomafana.'

'Well, somebody, for some reason, is spreading them.'

'Roy, you've worked here a long time, too. You know that rumors spontaneously generate. Did I ever tell you about the time back in 1980 when we tipped over a bus full of a BBC film crew? It was just an accident on slick mud. By mid-afternoon the whole town was sure we'd been attacked by bandits. They even had proof—a mark on the driver's temple. The fiendish bandits must have aimed their sling-stones so well they'd knocked out the driver of a moving car! The truth is the driver had a birthmark. Roy, you can't pin a sinister explanation on rumors in Madagascar.'

'Well, in my book somebody's out to get those kids.'

I didn't want to be talking to Roy. I wanted to go out and play. That is, I wanted awfully to be up in the hills watching lemurs—because the sun came out.

The evaluation team was assembled in the house where Pat and I stayed. Bob, Don and Hugh were wild to escape from the Hotel des Thermes—its sagging beds, cold showers and feeble lighting. Roy Hagen of ANGAP and Jasper of Water and Forests were with us only briefly, returning next day to Tana. We simply had to meet with those two and with Jean-Marc of AID, to be sure our basic documents all tallied, between USAIDS's version, ANGAP's version, the Water and Forests version. Predictably the documents didn't tally, so we worked through clause by clause. I was fascinated by Roy's views on Joe and Dai and

Dennis and the sinister rumor-mill... but oh what a drag fossilized documents can be.

I hadn't seen Dai or Dennis or Joe since C.J. dropped the axe.

So today, Friday, we were handcuffed to a meeting. Little by little the clouds lifted until the rainforest rose above Ranomafana town, bathed in sun. Sun here has a peculiarly golden tone, like sunset at noon. The folds of hills were deep green and black, covered in trees like tufted velvet. As the rain clouds lifted the forest itself began to steam. Wisps of mist rose from high valleys and hovered near the crests—white gauze veils over emerald velvet trees.

Even in the one muddy crossroads of Ranomafana town, birds started to sing. Up there above us in the forest, red-bellied lemurs would be uncurling from the branches where they had huddled against each other for warmth. Chocolate and white Milne-Edwards' sifakas would stretch and then bound 15 feet from tree to tree and climb the highest emergents to bask in the sun. The secretive bamboo lemurs might wake from midday sleep to come and nibble the cyanide-filled shoots of their daily meal.

And I was being paid to stay in a concrete house comparing clauses ... well, if we couldn't watch lemurs, maybe the sun would stay out for Saturday. We were planning all Saturday and Sunday to see Joe's and Dennis's work on the ground, so they could then go up to the Capitol in order to plead their case with AID. They still had hopes they could get a fair hearing—even though C.J. warned it would cut no ice with USAID.

The outlines of our report were already clear. Bob Brandstetter and the rest would say that the development side, in spite of all Joe's and Dennis's and Dai's good work, was failing in the impossible task it had been set. For all the nascent projects on tourism, forestry, understanding the villagers, the NC State crew had not convinced they were succeeding with people like the king's spokesman of the village of the fig tree—or 150 other villages around the Park. The naivety of the drafters of the ICDP idea condemned their work from the start.

Meanwhile, I argued for the burgeoning research agenda: science as it should be done.

Saturday.

The headquarters of the Ranomafana National Park project is now at the Ranomafana main—crossroads. The side street was actually paved wide enough to be a mini-square. Some civic-minded soul has erected two basketball backboards with rusty hoops, so half the street can be a playing court during hours when it doesn't rain. The square is lined with tiny wooden tin-roofed stalls, which offer cabbages, pineapples and balls of fried 'sweet bread,' Madagascar's version of doughnuts. Most of the stands are fringed with what look like gray and brown bead curtains. These separate into individual necklaces for sale to tourists. The necklaces are grey-white seeds and what seem to be lacquered golden-brown pine cones. They are not cones, but the reproductive heads of *Cycas thouarsii*. Cycads date from the coal age; trees which grew before bees invented flowers. One lyrical botanist calls *Cycas thouarsii* 'the coelacanth of the vegetable kingdom.' Imagine a necklace of coelacanth teeth!

Park headquarters is a concrete house shared with a carpenter's shop and saw mill. There are no fires, and it is bone-chillingly cold as rain pelts down outside. I hang my dripping jacket outside Joe Peters's office, but there is no temptation to take off my gray ski-sweater or wellington boots.

'I am really sorry to ask you for more of your time, Joe, at what I know is a really rough moment for you. But I wanted to know about those air photos. Also, I basically don't understand why conservation and biodiversity are split the way they are—to me, conservation is conservation of biodiversity.'

Joe looks just as white-faced as before, in the gloom of his office. He is wedged between a metal desk and a bookshelf of serried binders, side-lit grayly through the streaming windowpane. As he talks, he pulls out papers to illustrate his points. Does he spend days maintaining this superb documentation? In any case, I suppose this climate does not produce tropical tans!

'What I do is called conservation; what the scientists up in the research cabin do is called biodiversity. North Carolina State hired me to help the chief of party delimit Park boundaries, identify important or vulnerable areas of the park, and help in administration. I am also writing my thesis on ecotourism—oh, but you have already seen the survey I did on tourists to the Park.'

'Tell me about conserving the Park itself—what is your basic data?'

'Two years of nearly complete air photo coverage at 1 to 15,000. Here's the base map that shows the location of each photo. Here's a couple of photos—you see, they slot in just there on the map. 1990 is a little blurry, but on the 1991 series you can see every hut and trail. Look—there's a small landslip from an abandoned *tavy* field.'

'What has been done with these?'

'Not much so far. They were fundamental for delimiting actual Park boundaries. We had to move the boundaries inward because the air photos showed some villages inside the original planned limits. When Pat Wright and the others walked around the whole Park, explaining the Park idea to villagers, they went around the outside of a swamp—this lighter patch here—and just missed fifty-year-old villages—

these ones, on the Park side of the swamp. Air photos were invaluable for that. The air photos also are the basis for Bruce Johnson's soil sampling, which underlies Dennis's work. But the photos have not been used for an ecological map of the forest itself.'

'Will that give not just a spatial, but a historical understanding of the forest? I would not be surprised if some areas of Ranomafana's forest are old *tavy*. With today's population pressure people scrape *tavy* clear so it never regrows—but there could still be abandoned and regrown sites on the high hills.'

'Your guess is just right. High up in the Park there are archeological remains of old villages with fortified walls and ditches. These people aren't just fantasizing when they say the Park is ancestral land.'

'I think I do understand why this part of the work is called conservation. It's going to be so expensive that only the AID grant could cover it.'

'Sure. If it's USAID, it's conservation; if not it's just research.'

'Do you work with anyone else?'

'We used to work with a Catholic NGO. It did some of the social side. There were financial irregularities, though, so we cut them off. They are based in Fianarantsoa and they don't try to do much around here any more.'

'What else do you do as conservation?'

'Local conservation educators, based in villages. Sixty big yellow signs for the Park frontiers. *Procés-verbals*—that is, summonses. The local forester has served eleven criminal summonses since the project started to people doing *tavy* or taking timber from the Park.'

'Is there actually still timbering?'

'Well, we're down from ten to six forest exploiters, which is a start. Public enemy number one actually threatened people with guns. He was taking rosewood out of the park until we got a *procés-verbal* served on him in 1991. It was issued out of Fianarantsoa, the provincial capital, because various people in Tana were back-pedaling, and the local forester thought action was unwise. I can understand the local forester. He's only been here six years, since the last one got murdered.'

'Wait a minute—I was around here when that murder happened. He tried to take a short cut home from walking the boundaries with Pat. Didn't some men from a very remote village chop him in pieces and hide the pieces in the forest—and when the police came to ask questions a terrified village child admitted seeing it all? The story was that the villagers were either so frightened or so covetous that when a stranger appeared they just attacked him.'

'That's right—it was this village up here.' Joe Peters poked at a hamlet on the air-photo map. 'I've been to that village to see the people. They're all out of jail now. They're as nice a bunch of rural farmers as you could hope to meet. You know, there is remote and remote. It's only two days' walk up to there.'

'You're right. I also heard that the forester was very, very frightened long before it happened... It makes more sense to look for cause and effect than mysterious coincidences.'

'You might want to check up on public enemy number one.'

Sunday.

We have escaped from offices! We are going to a village! We will see Joe's ecotourism plans and a new school, and Dennis's agriculture. It is, of course, raining, but it may be going to rain for the rest of our lives.

The team has drawn a deep breath, and decided to go a little way into the rainforest—in the rain. How can Pat make them see what she sees—an enchanted forest with a dozen species of enchanting lemurs—when Hugh and Don and Bob's boots and binoculars and tempers are waterlogged?

Bob, Don, Hugh and I stair-step down the gorge-side trail. At the bottom the river rushes over a lip that is half-rapid and half-waterfall, swirling in foam under the foot-bridge. The footbridge regularly whirls into the gorge during cyclones. This third or fourth version is a great improvement over the first one I knew. It is still two trunks with wide-spaced cross-steps—but now there is a handrail. I still don't look down.

We climb to the deck of a Lincoln log cabin. Inside all researchers share communal meals, and outside they congregate to write notes or dry laundry under the porch tarpaulins. Most of the researchers are up in the forest—rain and mud and leeches do not daunt this crew. In fact, the more we ask, the more cheerfulness seems to steam off them, now and later. They are privileged to sleep in tents in the mud. They are privileged to work without hot showers or dry clothes. They are privileged to study in the forest. Don't we want to go look? Even a little?

All the rest of the team say no, not today, not even a little.

We push on up the road to the head of the escarpment, and now we are sitting in an icebox of a school.

The brand new school of Vohiparara isn't even dry yet. The smell of wet concrete plaster blends with the smells of village mud and cow dung and wet villagers. There was frost here last week, for the first time in living memory. Banana trees outside the glassless school windows rattle brown sails. They are seared brown as though fire had passed, not frost.

Bob, Hugh, Don, Jean-Marc and I are on the teachers' front bench, hugging ourselves to keep warm in our boots and pullovers. Twenty-five villagers sit wedged into children's bench and table sets in the one-roomed school. Few of them have shoes. I keep looking at the Mpanjaka's (the village king's) old arthritic toes on the concrete floor. Their hospitality consists in an inordinately long speech in flowery French by the village schoolteacher. The key to good manners is to sit in gelid immobility as though they wanted the speech to go on forever.

Vohiparara is a Betsileo village, lying on the main road just above its plunge over the escarpment lip, above the first thundering waterfall. Joe, his American length jackknifed into a child's desk, nods approvingly as the teacher explains that the Betsileo plow their rice fields and take from the forest the endemic river crayfish, which they sell by the road, and straw baskets of wild guavas, which they also sell.

They have plans, though. The park has brought them hope. They plan to set up a roadside restaurant for tourists, and a hotel of simple chalets. Their young people's group, led by a dynamic youth who has finished half of high school, has cut 5 km of nature trails for tourists that start at their village and join up with the research trails below.

The schoolteacher hits his stride. With a flourish he unrolls an architect's plan. 'Here are the tourist chalets: A-line roofs, pitches steep to shed rain. The site already chosen and agreed. Six months of lobbying done for the necessary permissions and the grant written for investment capital. We have even considered the water supply. We ourselves drink from the river: we are often sick. We have identified a clean spring only a few kilometers away which could be dammed to bring water to the tourist complex—and to the village that needs it so badly!'

We lunge at the opportunity to stand up and inspect the plans. At last there is a polite way to move.

Everyone shakes hands. We flip up our raincoat hoods; the schoolteacher dons a huge straw sombrero of fuchsia and green to shed the rain. We stepping-stone across the mud lake of the schoolyard, toward the road and proposed tourist site, a field criss-crossed by fallen trunks.

'You can't see the river,' I say, disappointed. 'Well maybe if you sited it facing that way, you can see a forested hill...' They stare at me in surprise. Apparently they had not thought that tourists would want the bungalows in touch with nature.

Don Stone and I drop back as we walk toward the bus, heads down in the rain, so voices don't carry, 'Have you read the health report on Vohiparara?' mutters Don. 'Some of these people set a world record for intestinal worms. We're in a world record village!'

'Those gruesome things in formalin at the museum? *Ascaris lumbricoides*: nematodes the size of earthworms?'

'That and a lot of other kinds—but *Ascaris lumbricoides* alone is 14 per person. Also infant mortality is highest here in the whole region. Average number of deaths per woman in Madagascar overall is 1.5 out of 6 pregnancies. Vohiparara families lose 2.6 kids out of 6 pregnancies. Do you fancy eating in their restaurant?'

'No wonder they want a clean water supply. That would help.'

'Even if they clean up the water, who here knows how to run a hotel? Even in the Hotel des Thermes we can't get a dry towel to go to the baths or any towel at all much bigger than a washcloth. How is Vohiparara village even going to deal with soap and towels? I think the whole thing is a pipe dream. At its harshest, I think these villagers have been misled into hoping for something that just won't happen. All their work on plans and permissions just makes it worse.'

We climbed into our waiting minibus, and waved. The youth leader waved back, and the Mpanjaka with his toes in the mud, and the schoolteacher engulfed in his pink and green sombrero.

A few hundred meters along the road, we stopped at three fields unlike others. Most rice fields here are waterlogged in winter—it keeps down weeds. A rice-growing landscape is laid out in little patchwork squares on human scale, and in winter each is a small placid mirror of the sky.[3] These fields were drained. They had high, geometrically straight field-dikes between them, and a 2-foot-deep channel of fast-flowing water down one side.

'Meet Madame Pauline.' Dennis' rapid-fire Hispano-English barreled through the introduction. 'Madame Pauline is the very best farmer in the region!'

Madame Pauline was short and stocky. Her hat brim brushed my shoulder. Her hat was so old and oily it almost matched her brown face. It was the best kind: its rickrack straw weaving so tight no raindrop penetrates. I have one too—a new pale gold one that I wear to summer weddings.

She indicated her fields with a gold-toothed grin. Stakes with labels divided them into lines—no fertilizer, urea only, complete fertilizer formula. She's had a year and a half of trials: the first year the crop wiped out in drought (drought? yes, it happened). The second year she harvested ten times her usual yield, going from half a ton to five tons per hectare.

She has eleven children, all grown up. Three still help her here. Where are the others? Oh, in Antananarivo, the capital, working or studying. Does she have family there? Oh yes, she comes from Tana, not Vohiparara. And how does she farm all that land? She hires help. The project has paid for laborers to do the experimental trials. In other years she would sell a cow to pay.

3. For this and more lyrical descriptions of rice, see R. Decary, *Souvenirs et croquis de la terre malgache* (1969).

I ask Dennis, 'Is this a fair trial, or does it just show other villagers that the rich get richer?'

'What do you want? You ask me for results in two years! The villagers are worse—they want results in one year, or right away. They can't afford to wait. I can't afford to educate the first model farmer for ten seasons—I need Madame Pauline! Dennis's arms semaphore and his shoulders shrug.

Madame shivers and pulls her wet *lamba* round her, as a gust of wind hits the back of her middle-aged legs, and the rain slides down her bare feet.

'Let's go!' says Don. 'I'm glad to see these people are human. I was beginning to think no one but me would ever start shivering.' He bows a courtly goodbye to Madame Pauline, who smiles from beneath her hat.

Dinner in the Hotel des Thermes follows French rules as to the number of courses. At least we have abandoned formality, to drag a table right in front of the evening fire in the colonial grandeur stone fireplace.

Hugh drawls, 'Did you know there are a 160 villages within 5 kilometers of the park? That's almost 30,000 people. These people are all getting their hopes up that the Park project will recompense them for losing their land. They're liable to get disappointed.'

Don cuts in crisper tones. 'I would be terrified if I worked on this project. The whole thing could just implode.'

'I don't understand. What do you mean?'

'We-ell.' (Hugh can sound awfully like a Florida cracker, when he isn't dropping reminiscence about Honduras, the Philippines, Indonesia, babirusa boars, and the downfall of the Mayan Empire. He told a delicious story about catching a citified trespasser picking mushrooms on his Florida land. He leaned over the fellow in his dungarees and best cracker accent, 'I'll bet you've seen lots-a shotguns in the back windas of pick-up trucks here-abouts.' 'Yes,' wavered the trespasser. 'Now I'll bet you think they're fur shootin' waaaild animals!' The trespasser high-tailed it back to the city.)

'We-ell. Dennis is doin' lots of good work. I'm impressed with his fishponds and that model farmer of his that got five tons per hectare of rice. Three villages have new schools that we get to watch inaugurated come Friday. But there's 157 villages that don't have schools. There's almost 30,000 people that gave up their land to the Park, and haven't got diddly-squat.'

'You have to start small—they have to begin with a few model farmers and pilot villages.'

'We-ell, yes. But what about just lettin' them all have some fertilizer?'

'Fertilizer?'

'Y' see, the phosphorous content of this soil is so low that things just don't grow. Now I'm a low-input man, myself. Other places I'm all for organic farming and recycling. But here, there's nothin' to recycle. You plow in plant residues and manure—it's all from plants and animals grown on this same low phosphorous soil. You never get past the limiting factor!...

'With some phosphate, and nitrogen-fixing plants, then they can increase total fertility by recycling, but as far as I'm concerned, the project design was flawed from the beginning. You can't possibly impact 30,000 people in three years. They didn't stand back and say, do we really mean to compensate these people for losing their land?'

'Its a time bomb,' reiterated Don. 'It looks as though Joe and Dennis are out. That means six months at a minimum before their successors are appointed, at least a year before work starts again. A new outfit never wants to build on its predecessors' ideas—they always start over. The local people may not get anything concrete for a year or more—and Joe and Dennis did a lot of good, even if we don't like it all. Hope is a terrible thing. This project has taken a terrible risk, building up people's hopes… I'd be scared to death if I worked here, that it would just implode.'

Wednesday.

Dai is at the door. She wants to see me. I don't want to see her. I have been avoiding her for two whole days while Joe and Dennis are in Tana pleading with USAID. I have been avoiding that vision of her naked gold legs on the motorbike, and her black hair blown back, which may be pure fiction for all I know.

I pull on my sodden jacket and follow her into the rain. We trudge to Main Street, to the Hotely Kanaka, not the Thermes where there will be evaluation team members. The Kanaka is a corrugated iron shack with holes for windows, wretched coffee, good vanilla tea. Dai and I have citronella—lemongrass tea. Same plant as the insect repellent, but it's lovely as a drink.

I look at Dai. She is indeed beautiful, in a strange way, like the witch or the princess of a fairy tale. She wears a long dark flowing skirt and has a dark raspberry-colored shawl looped round her neck. Why do we pink people think of pink as our color? Warm pink light reflects up under Dai's golden chin and cheekbones, a private silken fire in the blue, rain-spattered wind from the window hole.

'I don't know who else to talk to.' she says. 'You seem kinder than the rest. It's all so horrible. There are all these rumors going round—and they all seem to center on me.'

'Well, yes… I know what you mean.'

'You must believe me. You just must believe me. There is nothing at all between Dennis and me. We are close friends, but I haven't been having an affair with him. Really nothing. All these foul rumors must be meant to get Joe and Dennis thrown off the project—and we all have so much left to do here!'

Dai, the redoubtable Dai, is crying. Tears hang in the corners of her almond eyes.

'Look at it our way. Joe and Dennis have been here all along, slogging their guts out. So have I for that matter, though I only was a short-term consultant to the project and now I have my own Fulbright grant. There is no way people can discredit Joe and Dennis's work, so they do it this way, through me.'

'I can believe there are rumors. Rumors are quintessential Malagasy. The gold-digging one's a classic. That has been going around since Flacourt came here in the seventeenth century. But the ones about you and Dennis... What other enemies have you got? Has anyone propositioned you, for instance, that you turned down?'

Dai's eyes widened. 'No—no, not really.'

'I don't want names,' I said hastily. 'I just mean if you can think of anyone who is so jealous of you, or Joe, or Dennis, that they'd stir up trouble so bad it would reach USAID.'

There was a long pause while I listed enemies in my mind. People disgusted by Dennis's 'cat latrines' or Dai's video. People jealous of the model farmer Madame Pauline. The Catholic NGO ousted for mismanaging funds. The well-connected gun-toting rosewood magnate summonsed by the forest service. The forester whose predecessor was cut into pieces, who is forced to serve summons when before he could have sat back and taken a bribe in safety. The local reserve chief, his power usurped by the whites so the women on his staff are fairly safe from his harassment. Unnamed, shadowy bureaucrats in Tananarive (I know several names) who are losing power to ANGAP, the new

National Parks service, and who would love to see ANGAP's showpiece project come crashing down. Then there is Don's apocalyptic vision of the project itself imploding from hopes denied.

'No,' said Dai slowly, 'I don't think we have any enemies.'

'Dai, there's not much more I can say. I don't know the truth of any of this, and I probably never will. I have just one more question. The only eyewitness account in the whole story is the chief of USAID, who apparently...'

'Who came down here and saw me riding on a motorbike with Dennis, and said half my ass was showing. I was wearing a business suit. Come down to my house and I'll show you what I was wearing! Its just like a business suit!'

Later I told Dai's story to my daughter, who chided me that it probably was a business suit, only that when riding pillion on a bike, a skirt inevitably creeps higher and higher...

Friday.

Vahondrano, Vohiparara, Ambatolahy have their new schools inaugurated today. I stand in Ambatolahy, the village of the male stone, where a waterfall sings its way down a side gorge to the main river. The new school stands on a shelf above sparkling rapids. It is plastered pink and glows in the sun like Dai's scarf. A zebu cow, sacrificed for the inauguration, is blood red, liver-maroon, rib-cage white in the sun. Pat Wright is in a purple dress, laughing and making spellbinding speeches. The head of USAID is laughing in a brand new square straw chief's hat presented for the occasion. Joelina Ratsirarson is laughing. Long gone are the memories of his humiliation at the hands of his Napoleonic minister, when we all trekked across the States from St. Catherine's Island. Ratsirarson is now secretary general of Water and Forests, the top-ranking Malagasy of all those who have come from the capital to inaugurate Ranomafana's village schools. He is basking in others' adulation as they toast him in whisky.

The village band plays traditional cowhide drums, home-made guitar, a percussion xylophone and a rasp made of punctured condensed milk tins. The waterfall adds bass and all the forest birds an obbligato in the sunshine.

The head of USAID spots me. 'Isn't this a great project, hey? Haven't they accomplished lots in the time? Haven't they all got the right to be proud? Specially Pat Wright, of course. Never would have done it without her. Never would have started.'

'I wonder, could you tell me what's happening to Joe and Dai and Dennis?'

'Oh, they're out, and they won't be back. But hey, isn't it a great project?'

SIXTEEN

Real life and DreamWorks

By the end of the first five-year Environmental Action Plan, it was clear to USAID that the Integrated Conservation and Development schemes needed major overhaul. They should never have promised better livelihoods, let alone health care, to all the surrounding villages. They should never have drawn an artificial boundary of 5 km round each reserve to choose villages to be helped. They should, above all, not have waded in with almost no knowledge of village structure, even given the urgency (to Western eyes) of saving crucial zones of Malagasy biodiversity.

Meanwhile, development practitioners and anthropologists began to focus on the deficiencies of the whole system from the point of view of the rights of villagers. Joe Peters wrote up his own critique of the Park procedures. He concluded that the ICDP model of protected areas plus buffer zones, with attempts at economic development, did little to distribute wealth to poor people and everything to encourage the middlemen (and women). Nothing short of a New Deal type Civilian Conservation Corps would provide the cash and employment so desperately needed.[1] Paul Hanson turned

1. J. Peters, 'Local participation in the conservation of the Ranomafana National Park, Madagascar' (1997); J. Peters, 'Transforming the Integrated Conservation and Development Project approach: observations from the Ranomafana National Park

his studies of 'the village of the fig tree' into theoretical attempts to understand the ideological as well as economic conflicts. Hanson harks back to two formal speeches at the dedication of a new 'foundation stone' for the park—that is, a stele invoking the blessing of the ancestors and the unity of the participants. This was in 1994, a year after the ousting of North Carolina State. On that occasion Pat Wright spoke in Malagasy about her thanks to the people, but also the need to abandon slash-and-burn for the sake of their own forest. In return an elder made a flowery oration about the meaning of the forest to his people, identifying tombs and ancestral sacred sites deep within what had become a National Park, now barred to the descendants of those ancestors.[2]

Hanson quotes anthropologist Eva Keller: 'To tell the Malagasy farmers to preserve ... biodiversity by stopping their growth on the land—that is, by having fewer children and not creating more "land that enables life" and "land of the ancestors"—is not simply a request to change a certain mode of cultivation. Rather the conservationist program is an assault on one of the most fundamental values held by people in rural Madagascar: that is, the value of the growth of life through kinship and through one's roots in the land.'

The most telling account came from Janice Harper, a medical anthropologist who lived near Ranomafana in 1995. The American Park Project manager, by now leery of outside anthropologists, told Janice: 'If you cross Pat Wright, you're out of here.' Janice never actually met Pat Wright or found out if that was true. The manager made her life miserable from the first delays in research permits through the horrible

---

Project, Madagascar' (1998); J. Peters, 'Sharing national park entrance fees: forging new partnerships in Madagascar' (1998); J. Peters, 'Understanding conflicts between people and parks at Ranomafana, Madagascar' (1999).

2. P.W. Hanson, 'Engaging green governmentality through ritual' (2009); P.W. Hanson, 'Toward a more transformative participation in the conservation of Madagascar's natural resources' (2012).

day when the manager denied Janice transport to Tana while Janice's father was dying in America. Meanwhile the young researcher had chosen to work in a village where the death rate was exceptionally high even for the region.

It turned out that one village family had access to a little cash from selling vegetables to the Park personnel. That family systematically starved their neighbors through loans on others' land and calling in the loans, reducing the other families to sharecropping. The age-old differences between people of noble, commoner and slave descent masked the inequalities exacerbated by that trickle of income from the park. The poorest died of parasites, malaria, pneumonia on top of their malnutrition. *In extremis*, some trekked the four-hour trail over hills and streams to Ranomafana town in hope of medicine, but clinics would be closed and the simplest pharmacy pills beyond their means.

Administrators did not care how their cash distorted village life. Harper notes that the research scientists themselves had much more sympathy with villagers than the administrators did: they worked with their guides and depended on them. But, not surprisingly, she concluded that the ICDP system was deeply, even evilly flawed.[3]

In many other parts of Madagascar the academic backlash against conservation intensified. Christian Kull wrote a highly influential book, *Isle of Fire*, that challenged the whole picture of peasants as sullen, ignorant butchers of their environment.

Bush fires in the highlands were outlawed from the beginning of French colonialism, but no government has been able to suppress them. As much as a quarter of Madagascar burns every year—grass fires set to provide a green bite for

3. J. Harper, *Endangered Species: Health, Illness and Death among Madagascar's People of the Forest* (2002); J. Harper, 'The environment of environmentalism: turning the ethnographic lens on a conservation project' (2008).

the cattle as the grass re-sprouts. This is crucial for extensive livestock raising: the answer to the limiting factor of forage for much-prized zebu at the end of the dry season. Fires also clear the land of invasive bramble thickets. Most fires burn unchecked across the hillsides, but when villagers wish to save houses or cropped fields or a sacred forest they do try to control the flames. Burning is so traditional and so ingrained that the alumni of a school or university are called 'well-burned land.'

Kull admits that slash-and-burn *tavy* in the forests does destroy irreplaceable natural habitat, at least in modern conditions. Even there the villagers are not stupid or ignorant: they know exactly why they do what they do. In rainforests north of Ranomafana anthropologists championed even *tavy* farmers' viewpoints, speaking for people at the bottom of the hierarchy who could not otherwise be heard by the powerful.[4]

As USAID realized the limitations of the ICDPs, for the second round of the EAP they turned to the 'landscape approach.' Lisa Gaylord, indomitable administrator of USAID's environment program, chaired a meeting in Washington in 1995 to explain the changes. No longer focused just on Ranomafana and similar ICDPs, the new program would embrace the whole corridors, including the one between Ranomafana and Andringitra, the dramatic National Park which holds Madagascar's highest mountain. The semi-forested corridor, full of people, is 150 km long. I cynically thought during that meeting that Lisa was making a desperate attempt to

4. E.g. C.A. Kull, 'The evolution of conservation efforts in Madagascar' (1996); C. Kull, *Isle of Fire: The Political Ecology of Landscape Burning in Madagascar* (2004); J. Pollini, 'Slash-and-burn cultivation and deforestation in the Malagasy rainforests: representations and realities' (2007), p. 776; E. Keller, 'The banana plant and the moon: Conservation and the Malagasy ethos of life in Masoala, Madagascar' (2008); J. Pollini, 'The difficult reconciliation of conservation and development objectives: the case of the Malagasy Environmental Action Plan' (2011); G.M. Sodikoff, 'Totem and Taboo reconsidered: endangered species and moral practice in Madagascar (Mananara)' (2012); G.M. Sodikoff, *Forest and Labor in Madagascar: From Colonial Concession to Global Biosphere* (2012).

deal with far bigger problems while putting in less money. However, as Karen Freudenberger explains in her magisterial review of the USAID conservation efforts, it was actually targeting the real problems with effective point interventions. Instead of promising general livelihood improvements to 150 villages within a 5 km radius of Ranomafana Park, the Landscape Development Initiative (LDI) would analyze the constraints for villagers and attempt to target these.[5]

Mark and Karen Freudenberger headed up the LDI for the great parks of Ranomafana and Andringitra and the forest corridor between them. They focused on a program of providing fruit trees that the villagers wanted. This should allow permanent use of cleared land. However, they found that, at least on the uphill Betsileo side of the corridor, it was actually the richer villagers who could afford to send their offspring and nephews out to clear the forest; the poorest were too close to the margin of life on their tiny plots to even try. Improving livelihoods might actually lead to more forest destruction, not less.

Even more telling, though, was that they realized the importance of the rickety rail line from Fianarantsoa town down to the coast, which traverses the forest corridor. This is a railway tourist's delight: a 1930s' narrow-gauge line with antique carriages and engines. It snakes down the escarpment, at one point curling under itself in a full circle. It's always breaking down, of course. When it runs it is the lifeline for the farmers. It stops every twenty minutes to collect the local produce of bananas, oranges and coffee. The main farmer constraint is not land or labor, but the possibility of a link to markets outside the region. After cyclones shut the line for months of repair, the farmers had no choice but to return to

---

5. K. Freudenberger, *Paradise Lost? Lessons from 25 Years of USAID Environment Programs in Madagascar* (2010).

forest-clearing for rice and manioc just to get something to eat.[6]

The landscape approach pioneered by Lisa Gaylord of course had its own problems. Those of poverty, lack of unclaimed land, impoverished soils and spiraling demographic growth were just too big to fight. Karen Freudenberger calls her review of the whole program *Paradise Lost?* She concludes that for everyone's sake one must keep trying—but she is not very hopeful.

As five-year cycles passed, USAID changed policies when it seemed necessary. Health-care attempts were dropped from the environmental program in 2001. A host of small NGOs have made point interventions, funded by USAID and many other donors, but under the budgets for humanitarian aid instead. The gaps are slowly being filled.

The Park itself has a splendid center. It is called Centre ValBio: Center for Valuing Biodiversity. Pat writes:

> Centre ValBio has training for Malagasies, including computer training and molecular biology, as well as being the hub of our health team that visits remote villages and our Unicef connecting classrooms program and our medicinal plants and reforestation programs. We work with the local villages training them in new markets for handicrafts, construction trades, tourism, computers, etc. It is a Conservation Hub where the people around the park and the biologists who study the biodiversity in the park can work together. The park, including the income from 26 tourist hotels, the tourist guides, park entrance fees (30,000 tourists a year in 2013), the salaries with fringes of our staff of 85 local people brings over 2 million dollars a year in cash economy to the area around Ranomafana. In addition we have just begun a new health program that will upgrade

---

6. M.S. Freudenberger and K. Freudenberger, 'Contradictions in agricultural intensification and improved natural resource management: issues in the Fianarantsoa forest corridor of Madagascar' (2002).

> the local basic health centers and the district hospital which adds more millions of dollars a year from my donors. Centre ValBio is a breathing center for training and for science and for sustainable development all in one package. It takes a long time to do conservation that lasts, and every day I hold my breath and fear it will all slip away into oblivion, but we have a great team fighting hard to make this long-term conservation project last far into the future. For the lemurs and wildlife and people together.

I asked Pat what news of Ambodiviavy, the village of the fig tree?

> Zanaka [became] Ampanjaka [king] before he died in 2008. Our health team visits Ambodiaviavy once a month since this village is part of our Health and Hygiene program for the past twenty years. We asked Hery, the head of our Health and Hygiene team to describe the improvements in Ambodiaviavy. Hery reports that now Ambodiaviavy has a water fountain (pump) in the village and they no longer have to walk a kilometer to get water from the river. The houses now have tin roofs and some have tile roofs. This keeps the rats at bay. There are at least 10 latrine toilets in the village. Paul Hansen's house used to be a school before becoming a place for the Community Health Volunteer to check kids' health. Now Ambodiaviavy has a well-built school which is located in Ankevohevo, just down the road. This year 3 kids out of 7 passed the first exam (CEPE) of primary school. One thing I have learned is that any kind of development takes a long time, but with patience and training the lives of the people do improve. I can send you a photo of the new Ampanjaka taken at our inauguration of NamanaBe Hall,[7] and a portrait of him by Pierre Men, which hangs over the front desk… Both Loret and Emile are doing very well and I will have dinner with Russ tonight in Tana. Thank you for writing all this history

7. NamanaBe Hall: the new molecular biology building at Ranomafana.

up. It was such a great trip down memory lane. What incredible lives we have lived and how lucky we have been.[8]

Perhaps after the high hopes of the early 1990s, one water fountain, ten latrines and three kids graduating from primary school in a village near Park headquarters seem to be slow progress, but at least it is happening.

There have still been many gaps over the years. That was clear when I returned with Pat Wright and the mega-donors of the *National Geographic* Research and Exploration Committee—including Roger Enrico, chairman of the board of DreamWorks Animation, who had just made the cartoon film *Madagascar.*[9]

Saturday, Dec. 12, 2004. To meet Jeffrey Katzenberg, CEO of DreamWorks, in Berenty.

Russ Mittermeier—in inevitable tattered field vest and fraying shorts—welcomes me. Jeffrey Katzenberg is amazingly ordinary in looks and manner—a middle-aged, trim, bald man in white T-shirt and grey shorts, with totally forthright and low-key manner. The old primate rule: secure dominants never have to act dominant. Others just get out of their way.

I ask right away about *Madagascar.* Katzenberg launches straight into the plot. Four animals are friends in the Central Park Zoo: Alex the lion, Marty the zebra, a comic giraffe, and a hippopotamus who is chic and petite and sassy although she's a hippo, and has more balls than the other three put together. Katzenberg grew up in New York. He was fascinated (like all us New York kids) by the Central Park Zoo—but even then he fantasized about what would happen if the animals broke for freedom.

8. Patricia Wright to Alison Jolly, emails, July 25, 2013; January 6, 2014; January 25, 2014.

9. E. Darnell and T. McGrath, *Madagascar* (2005).

Marty the zebra has a midlife crisis as he approaches his tenth birthday. He blows out his birthday candles; he wishes he were in the wild. The others are appalled. All four were born in the zoo. They are New Yorkers. Marty heads to Grand Central (he has heard that you can reach Connecticut from there, and that there is some wild there). Eventually, they wind up in Madagascar where the King of the Lemurs greets them with much song and dance. The real turning point, though, is that Alex, the lion, gets hungry and finds himself with his jaws closed around the zebra's flank. (Of course in the zoo Alex never knew where steaks came from.) In the end nurture prevails over nature, and friendship over all.

I like the point that nature is not kind, and especially the scene where the friends rescue a baby duckling, which is at once gobbled up by a gigantic crocodile. We need nurture. But I do rather sourly point out that lemurs are female-dominant, so it should be a queen, not a king. Katzenberg says, 'That boat has already left.'

He adds disingenuously, 'This will be very big for Madagascar. I made a film about Africa a few years ago. It was called *The Lion King*. It hugely boosted tourism to Africa. I think that *Madagascar* is a good film, a bit like *The Lion King*. It opens everywhere simultaneously in May—the US, Europe, Japan. Madagascar is going to find every single airline seat taken this summer, and every hotel room booked. They'd better be prepared though—get their websites up and their tourism streamlined, because they'll never have this much publicity again.'

I tried to press him for some more immediate cash for Madagascar, but he deflected my urging. It turns out that he was saving the announcement to make to and with President Ravalomanana that DreamWorks would donate $500,000 for the promotion of Madagascar's tourism!

The next year, in 2005, the *National Geographic*'s Research and Exploration Council toured Madagascar. Patricia Wright, by now a leading member of the *Geographic* team, kindly swept me along with them to Ranomafana. Two of the Council members were Roger and Rosemary Enrico.

April 30th, 2005—Saturday, Ranomafana.
Roger Enrico retired three or four years ago as head of PepsiCo 'in order to have fun.' Then his old friend Jeffrey Katzenberg called up, said DreamWorks Animation was going public, and wanted Roger to be chief executive on the management side. Roger has always been fascinated by animated films, but demurred. He took on being a one-third-time chairman of the board instead. He and Katzenberg toured America together raising money for *Madagascar* to the point where they could finish each other's sentences.

Now he was with the other major donors of the National Geographic Society, junketing to Madagascar in a private plane with five obsequious stewards and tour guides plus five scientists to give them an absolutely top-level tour.

In our six-car convoy down to Ranomafana, Pat Wright and I were miked up! I gave a spiel into my little curved mouthpiece that duly rebounded into everyone's lightweight plastic earpieces in the other cars. I told them about Betsileo and Tanala, the 'good,' industrious, ambitious uphill paddy-rice people, and the 'bad' downhill, shifting cultivators in the forest. These categories are somewhat superficial in kinship terms: you can marry across, and change, tribe with your changed way of life, although Dai Peters described the overall societies well. It still matters much more within each 'community' if you are descended from nobles or slaves, and if, as the village reckons, you are currently rich or poor, even if the whole village is invisibly poor to Western eyes.[10]

10. J. Harper, *Endangered Species: Health, Illness and Death among Madagascar's People*

We stopped at a village; not Vohiparara, the world-record parasite village, but I think the one where we went in 1993 to meet Mme Pauline, the progressive demonstration farmer. They are actually forging iron using the traditional two-cylinder bellows. They made a shovel blade as we watched. I am sure that none of them would have believed how recently Mr. Windlass, our neighbor in Lewes, England, retired from working at his forge hammering out horseshoes.

Peaked highland houses, some of them brick with little sagging second-floor porches, but many the ancient mud style where the cows lived downstairs. 'To keep them warm?' asked someone. They were quite shocked when I said 'for fear of bandits'—and that the fear of bandits on the road stopped the kids walking the 5 km to school when Richard and I were last here.

All so picturesque—the two tiny girls pounding rice in a mortar by the road, the gabled houses and clanging forge—and all so very, very muddy and dirty. Here's another stereotype: I find Betsileo dispiriting. They work hard and they don't laugh much. Like Dai Peters, give me the gutsy coastal humor any time, even in Ranomafana where the coast starts 10 km farther downhill.

Downhill! The river tipped over the first waterfall. Tree ferns! Greenery! White flowers of ginger sprouting in every free patch along the road, and the first glimpses of the granite cliffs of the escarpment.

Over Pat's radio we heard her actually crying with joy to be back.

Orchids in clay pots (not tree fern pots any more), a terrace with view of the plunging forest-clad hilltops—and an awful lot of agricultural land right up to the hard line

---

of the Forest (2002).

where Park protection starts. The landscape still seems divided into two opposing camps: us against them. And them are the people who live here.

We geared up for a night walk right away. Total haul for the night—one absolutely tiny chameleon,[11] being endlessly sucked by a mosquito while people take close-ups. Pat suddenly wonders if reptiles have their own malarias, as birds and lemurs do. The nasuta makes no move to dislodge the mosquito till the insect flies away. They were about the same size relation as horse and jockey. At least the trail was dry! Hardly slippery at all! And of us all, only Roger Enrico, whose group climbed much farther than we did, picked up a single leech.

Sunday, May 1, 2005.

It is sunny! Lucky for Pat. Walk walk walk, up steep slopes. Pat's sifaka come to meet her! Or, rather, group E comes, way out of its territory, heading for guavas. Pat points out they are nervous and won't stay long, just a matter of time till the real owners chase them out. They do stay long enough for photos. Gorgeous chocolate and white, that mournful sifaka peering face, fur deep enough to lose a hairbrush in against the habitual clammy wet climate.

Climb climb climb. Belvedere, built by the BBC. Rolling rainforest hills in shades of green and the granite cliff faces in shades of Oh-my-God-look-at-that!

We started down. A pair of avahi are apparently sleeping ahead. Guide Rodin was so eager to show them to us that he took a short cut. Straight down through brush on a trackers' trail for twenty minutes or so. Into a marsh at the bottom, where Gail Roski lost a shoe in the mud and retrieved it the worse for wear. (The first Wikipedia entry for the Roskis is the Forbes Rich List.)

---

11. *Columna nasuta.*

And up again—this is not a tourist path.

The avahi curled up in near-invisibility, at least to cameras. The muddy, scratched, panting millionaires, though, were delighted. 'At last we realize what it takes to be a real primatologist!'

Lunch waiting, and a day of full sun! It couldn't have been a more judiciously planned sample of 'real primatology.' Bravo Rodin—the high point of the trip for many of the gang.

Then to ValBio. A stone-built building, every stone laid by hand, every bit thought about, including the all-important tack rooms for shedding muddy boots and rain gear. Much of the architectural planning was done by one of Pat's brothers from her extraordinary family of six kids brought up in poverty who are all profs, lawyers and architects. The lab is cantilevered over the gorge below the waterfall—a view from every balcony. Brilliant. Bravo Pat. And bravo the Finnish government, which paid for much of it.

ValBio is a far cry from village architecture, but what a difference it makes not only to local outreach but to future Malagasy biologists! Ranomafana has so far produced 19 Ph.D.s, 88 D.E.A.s (advanced Master's) and now has 65 local researchers in residence. Five of the Malagasy Ph.D.s got them from the USA, and all are employed in Madagascar. And Pat's own output to knowledge of the forest is prodigious.[12]

We assembled in the main room, with its big fireplace—not needed today in summer. Presentations in English by

---

12. P.C. Wright and B.A. Andriamihaja 'Making a rain forest national park work in Madagascar: Ranomafana National Park and its long-term research commitment' (2002); P.C. Wright and B.A. Andriamihaja, 'The conservation value of long-term research: a case study of the Parc National de Ranomafana' (2003). For fundamental research, see S. Goodman and J.M. Benstead, *The Natural History of Madagascar* (2003), chs 13 and 20.

the Malagasy heads of departments. Newsletters, education, research, and a major scheme for exploration of possible medicinal plants, with half the profits promised to the village communities. A lovely thesis on the endangered pitta-like ground roller, *Atelornis pittoides*. It's colored like an explosion in a paint factory, and nests in burrows. But I am most interested by young doctor Joel Ratefinjanahary. I have read so much of the continuing desperate need for health care, and how little the Park itself can do about it.[13] I want to find out if this is still true. Roger and I sit with Joel at lunch.

Joel monitors village health in four-day research visits per site, a side project to the biodiversity monitoring. First they visit the Mpanjaka, the village king, then for two days do household interviews, then on the fourth day have a focus group on community health, followed by a communal meal. The village gets out of it a gift of soap, impregnated mosquito nets and the meal. The goal is to understand how and why people switch between agricultural modes: paddy and *tavy*. Some 50 percent of adults have no education, over 70 percent of the population are under 15 years, most are underweight. Weight changes markedly with season. Besides soap and mosquito nets, they give people toothbrushes! Not v. useful.

The three-person team do vaccinations, given free by Unicef. There is no health treatment in villages. Joel does first aid, but strictly he should have a permit even for that. He has no medicines to give out. People know he is a doctor who would be able to treat them, but he has no permit and no medicines—almost as awful for him as for them. People die of diarrhea, malaria and lung diseases.

In Phase I of the EAP there were two doctors and nurses circulating through villages, though admittedly mostly

13. Harper, *Endangered Species*; Wright and Andriamihaja, 'The conservation value of long-term research.'

focused on birth control. Funded to $150,000 for three years, 1992–97. In EAP II nothing: USAID stopped its budget for health. The development group working with the park now is ERI,[14] which is interested in rural development and dams, not health. Cornell left a while ago to concentrate on Lake Alaotra and the country's rice basket.

And, yes, the peasants are still saying that the Park stole their forest land.

At which point, Roger Enrico, who had said absolutely nothing throughout lunch, leaned over and asked, 'Roughly how much extra do you need for health care?

Joel thought rapidly. 'Perhaps $25,000 a year?'

Roger said, 'Give me a proposal, and I will pledge $25,000 per year for five years. That gives you your continuity. If it is going well, renewable for another five. Lever it any way you can, such as Unicef's essential drugs.'

Joel, open mouthed, asked me quietly afterward, 'Is this man for real?'

'Yup,' said I. 'If he likes your proposal, he can just write the check.'

He did.

14. Eco-Regional Initiatives, based at the University of Michigan.

SEVENTEEN

President Ratsiraka

Didier Ratsiraka was appointed as president of Madagascar in 1975 after a tumultuous period of student strikes, interim presidents and the assassination of his predecessor. He ruled until 1993. Ratsiraka originally championed Malagasy socialism, nationalizing many businesses. The disintegrating economy, the decision to accept foreign loans, falling commodity prices and international oil shocks brought the bankrupt government under the sway of the World Bank and the IMF by 1981.[1] The neoliberal policies of structural adjustment, which would today be called 'austerity,' brought in a thirty-year period of practically no economic growth. Civil service salaries were frozen; corruption at all levels became a way of life.

In 1991 an uprising in the capital forced Ratsiraka to give up power after his two army helicopters threw hand grenades onto a peaceful protest march toward his palace. The Forces Vives government took over under President Albert Zafy in 1993. However, economic decline and corruption continued. In 1996, Zafy was impeached by the National Assembly for

---

1. For an inside account by Léon Rajaobelina, then governor of the Bank of Madagascar, see A. Jolly, *Lords and Lemurs: Mad Scientists, Kings with Spears, and the Survival of Diversity in Madagascar* (2004).

incompetence, including his attempt to circumvent the power of the World Bank by seeking 'parallel financing.'

Letter to Richard, July 1995, Antananarivo–Lichtenstein. Great dinner with the Metcalfs, where Ros Metcalf diagnosed instantly 'you need a book' and pressed the 1,497 pages of Vikram Seth's *A Suitable Boy* into my hands.[2]

Peter Metcalf told me the whole soap opera of parallel financing. The new Forces Vives government has decided to escape from the structural adjustment of the World Bank and the IMF. Opportunely, a prince of Lichtenstein turned up, with some exceedingly dubious cronies. Apparently Lichtenstein has a plethora of princes, and this one was perfectly genuine—except as regards finance. The prince and co. promised to send a shipload of rice because they understood well the plight of the hungry Malagasy. The government trumpeted that they had found a source of 'parallel financing' which would free them from the Bank and the IMF.

You can imagine the resulting debacle, when in fact a shipload of Pakistani rice arrived—but payment was demanded on the spot, not deferred debt like the Bank.

Peter remarked, 'The Malagasy may well think of Lichtenstein as one of the world's great rice producers.'

After the fall of President Zafy, what followed was a democratic election. In the first round there were twenty-two candidates, almost all from Tana, with tiny splinter parties as background. No one in the country at large had heard of them. The second round came down to two candidates

2. Peter Metcalf, resident representative of the United Nations Development Program, and Rosemary Metcalf, lecturer in English Literature at the University of Antananarivo.

whose names people knew. However, though the constitution provided for impeaching a president, no one had thought to rule that he shouldn't run for re-election. The two candidates, then, were ex-President Zafy, just impeached, and ex-President Ratsiraka, ousted by popular uprising.

Ratsiraka won, to the moderate relief of the foreign community. As the American ambassador remarked, 'A better president is one that when you bribe him, he stays bribed.'

Patricia Wright and I met President Ratsiraka in a hotel in Washington, where she lobbied him for the success of Ranomafana National Park, and me for a conference of the International Primatological Society.

March 13, 1997. Madison Hotel, Washington DC.

Tall and much thinner than I expected, and not so old. (I suddenly think: five years younger than me!) His picture as a 30-year-old used to look proudly down from every official wall in Madagascar. He was so handsome then, roundish face, very black eyebrows, the military bearing which so often turns plump later on, photographed as from the upward gaze of a suppliant begging protection. The photographs in banks and post offices and offices grew dusty and sun-bleached over the seventeen years of his presidency, the red and blue naval uniform dimmed, epaulets tarnished with the verdigris of disillusion.

It now seemed a different man, thin, brisk, above all alive, with eyes that flashed question and answer. His right hand punctuated the air in its dark suit-sleeve, gleaming gold wristwatch, and heavy, classic, Antandroy silver bracelet. With all the evil things that have been said of him, nobody ever said he was stupid.

He sat on a brocade sofa with Mme Vaohitra at his left, wife of Barthélémy Vaohitra of WWF. Mme Vaohitra is the new minister of the environment. Pat Wright and I perched on armchairs, wound up tensely like—sifaka? Or do I mean fleas?—about to take off straight up in the air.

I opened with describing the International Primatological Society's 1998 Congress—saying it should become a major statement of Madagascar's importance to the environment and in the world. Not just primatology, but conservation as a whole. The Malagasy organizing committee want to call it 'La responsabilité de l'homme pour sa propre survie a travers la Conservation de la Biodiversité.'

He actually listened—and said that, though long, the title was good. That indeed primatology alone would be far too limiting. (I seem to meet lots of people who don't even pick up on the word 'primatology,' even if briefed beforehand, but he did.) He was pleased, too, I also asked for his high patronage.

I raised the question of refurbishment of the University before the Congress. He said, 'You find the money!' (They estimate $250,000.) (I wouldn't have said anything, except that Leon Rajaobelina told Pat we must, twice, and Mme Vaohita told me early yesterday to bring it up.) I demurred, and the president said, what about using the new indoor stadium. Mme Vaohita explained that scientific congresses need seminar rooms, not just a central hall. However, the president said, you need a minimum of decency. At least the toilets have to work! We changed the subject—but Mme V. whispered to me later that yesterday the Ministry of Finance had OK'd the refurbishment!

He started to talk about the richness of Madagascar, especially its *pharmaceutique naturelle*, and the wound-healing product Madecassol. Then he said that lemurs were reputed to confer prolonged youth! That more than one head of state had begged a present of a couple of lemurs from him—he wouldn't say who—and indeed those who had a couple of lemurs had lived very long lives. (I thought

Deng! Deng! Oh Lord!...) 'After all,' said the president, 'that is every man's dream, is it not?'

I sort of fumbled for words. Mme Vaohitra said, perhaps it is something to do with the glands they use to mark their territory. I said something silly about I hope it keeps people mentally young to have a couple of beautiful lemurs, not some chemical effect.

A nightmare: if lemurs were ever thought to be 'rejuvenating' it would take about two years for the Chinese to clear them off the face of the earth.

But President Ratsiraka's first act on taking power again this month was to declare the Masoala National Park. The last great lowland rainforest, where I spent a week in a hut in the rain in 1975, the year Ratsiraka first became president, and a few days in 1985 with Martial Ridy, *tavy* cutter. It is now the pride of the CARE NBO and of the New York Zoological Society (aka Wildlife Conservation Society). The previous Zafy government was dickering with a piratical Malaysian logging company to abandon the Masoala to them, and enrich their personal bank accounts. Also perhaps to abandon the rest of the forest. WWF Malaysia blew the whistle. Ratsiraka's immediate action was to throw out the Malaysian firm's proposals.

OK. He knows where the higher stakes are—his almost next act has been to come to Washington and sign the structural adjustment credit with the World Bank. But he is acting like a real president, not a pirate.

Pat congratulated him on gazetting the Masoala Park, and mentioned Ranomafana. He said, 'The place with the hot springs?'

'Yes'

'What has it?'

'I'm sorry—I don't know quite what you mean.'

'What has it?'

'Well I…'

'What lemurs has it?'

'Oh! A dozen kinds! Including the golden bamboo lemur, which was discovered only in 1986.'

'Pat, don't say it was discovered—Mr. President, Dr. Patricia Wright discovered it.'

'Well, yes, I did. The golden bamboo lemur only exists in Ranomafana, and perhaps Andringitra.'

'Do you have a picture of it?'

Pat opened Noel Rowe's lovely compendium of living primates, and found his picture of *Hapalemur aureus* chewing its cyanide-laced bamboo.

'How do you say it?'

'Hapalemur aureus.'

'Like ear [*oreille*]?'

'No, like gold.'

'Ah: alpha, upsilon, romeo, epsilon, upsilon, sigma.'

(He was, of course, a frigate captain, the highest ranking Malagasy naval officer, in 1975—all the other officers were French.)

'How many are there?'

'We don't know—only that it is rare.'

'What are you doing to prevent inbreeding, then, if there are so few?'

'It is a worry. Fortunately Ranomafana is a very large park.'

'I see—Really, you can only visit it by helicopter,' said the president.

I found myself grinning broadly—a wholly nervous primate fear-grin. To me, I think only of the presidential helicopters throwing hand-grenades on the crowd marching toward his palace in 1991, and the reputed fifty deaths, including women and children. After the killing people murmured that such a murderer could never go on being president. Of course that just shows how gentle a country Madagascar is. A mere fifty deaths would go unnoticed in most of Africa.

He didn't pause—of course, he traveled largely by helicopter as president. He went on, 'If you are taking delegates to your congress to Ranomafana, there is no airfield, so you have only the choice of road or helicopter.'

'We will have a big ceremony there this May, as well, for the handing over of all administration to ANGAP.'

But he did not know what ANGAP was. Mme. V. had to brief him about the National Parks and Reserves administration, set up so earnestly under the first environment plan to get those areas out of the clutches of the forestry department. The other meeting Pat and I were attending in Washington was all about the transfer of real power to ANGAP, under the second five-year tranche of the Environment Plan—power taken away from the foreign NGOS who've administered them so far. But the foreigners had $80 million over five years, while ANGAP only gets $40 million for the next five years, with a broader mandate. We will see if they can make it work.

The president then began to talk of his goal—a 'République humaniste–ecologique.' And 'Perhaps it is a dream, but as I said in my inaugural address, the fauna and flora of Madagascar should be declared a world heritage, just like Antarctica. We already have one World Heritage Park, in the west—Bemaraha—but all the fauna and flora should become a world heritage! Then at last we could finally stop the fires!'

Maybe, just maybe we will see the reprise of Radama I's ten years of progress in the early nineteenth century? Maybe we will recover from your statistics, Richard, that Madagascar has declined more in income since 1960 than any other country in the world, that its school enrollments have halved, worse than any other country, that by percentage of people receiving less than $1/day—72 percent—Madagascar is the poorest country of all?

Times have changed, the president has not changed but come back again; but the president as a person seems to have changed—maybe. It will be a wild ride until the country shuts again.

EIGHTEEN

Madame Berthe was dancing

Ranomafana's successes in research stand in glorious contrast to the surrounding poverty, not just of the region but of most of Madagascar. Pat Wright has found funds to build not one but two research buildings between the road and the waterfall. London has its Shard, New York the Freedom Tower; Madagascar has ValBio and the even newer Molecular Biology Lab jutting out their angled balconies at canopy height toward the forest across the ravine. All right, they are smaller than the Shard, but even more spectacular in their context.

This year (2013) there was an International Congress of Prosimian Biology, meeting at ValBio in Ranomafana. But I will conclude Part 3 with an earlier congress, the 1998 Congress of Primatology, an intermediate stage in the research development of Madagascar. I hope it captures the exponential growth of Malagasy environmental research by Malagasy as well as foreigners, and the excitement of every biologist privileged to work in Ranomafana's forest.

August 3, 1998.

'Is it going to rain like this in Ranomafana?' asked Richard.

'You betcha.'

Richard and I huddled under the awnings of the Groove Box Restaurant of Antananarivo, across the street from Tsimbazaza Zoo, watching rain sluice onto the street and splash right back up onto primatologists from five continents. We forayed out to find taxis, but there weren't any. So we all retreated inside, begging coffee and tea.

The wet, chilly crowd included Jatna Supriatna from Jakarta, organizer of the International Primatological Congress in Bali four years ago. Jatna invented the idea of a pre-congress meeting for Third World primatologists, to be held before the main Congress. In Bali, Jatna spirited off the scientists from developing countries to an actual nature reserve in the west of Bali. Bali isn't just gamelan gongs and temples and tourist hotels and Australian surf-boarders. There is also a wild place in the mountains. The earnest naturalists attended lectures and laid out transects and watched a demonstration of darting long-tailed macaque monkeys. Some even spotted the ultra-rare Bali myna in the wild, a snow-white bird with an erectile crest and sapphire eyes.

So, thanks to Jatna's example, we were all here four years later, trying to make friends in a motley gang: Ernesto Rodríguez-Luna from Veracruz, conservation secretary of the International Primatological Society; the distinguished, silent chief of Jakarta's Indonesian Forestry Department; Camilla and Esmeralda, eager undergraduates from Mexico; student Kaberi Kar Gupta from India. Kaberi wore a T-shirt with a hand-painted slender loris at about fifty times life-size ramping across her diminutive frame, its round eyes much bigger than her breasts. That's identifying with your study species for you... I tried conversation with Dr. Li, from China, but was baffled by his account of twelve years studying Chíbechan macaques, pronounced in descending notes like a xylophone. I muttered Chíbechan, Chíbechan,

Chíbechan, thinking I must recognize the species soon, until the word suddenly morphed into 'Tibetan.'

When the heavens opened at least we all had the prospect of adventures to start the next day.

August 4, 1998.

Off to Ranomafana! A cheery morning of leaping into the hired cross-country bus at 7 a.m., to take the merry multi-lingual group to our tents at nightfall. Well, not exactly 7 a.m., more like 8 by the time the luggage was lashed on top. Well, not exactly 8, more like 9 by the time the drivers thought of going to get diesel, and people delivered last-minute messages, and someone from Ghana was lost but the Ugandan had landed at 3 a.m. and merely overslept. Well, we might not be quite going to Ranomafana, because Patricia Wright had left the day before. She radioed back that the short-cut road was impassible with mud thigh-deep from thirteen consecutive days of rain. Anyway, we all climbed in and started off for somewhere.

The World Bank has fixed the tarmac, hooray! No more potholes on the highway south! But no straightening either, so a bus can wiggle even faster, slowing only for one-lane, one-way bridges, where carcasses of trucks in the rivers show what happens to those who did not slow.

We stopped at tiny roadside hotelys for violently black coffee and *mofo-mamy*, or sweet bread, the Malagasy roadside version of donut holes. We stopped at Antsirabé for lunch, and walked off to admire the restored grandeur of the 1900 Grand Hotel des Thermes, with its pinnacles and porches and the sweep of a palatial stairway of solid rosewood. We piled out to admire the woodcarvings of Ambositra, where Richard fell in love with a hand-inlaid wooden game cabinet with trays for backgammon and checkers and even scrabble, and chessmen in the form of plateau kings and queens holding their umbrellas, every pawn armed with shield and spear.

At last it was decided: no camping in Ranomafana tonight. Instead we holed up at the Sofia Hotel in Fianarantsoa. Patricia Wright herself materialized in a winged chariot—the mudded Land Rover which braved the road we did not take. Her black hair swung as usual; her delighted grin showed the crooked teeth.

'It's a great idea to have a night in dry hotel beds, isn't it!' (Let us not ask what thirty extra hotel nights does to the grant budget.) 'And just before dark,' beamed Patricia, 'I do believe I saw a break in the clouds to the east! Sun tomorrow!' Pat would never have created a National Park if she were not an optimist. Certainly not at Ranomafana. She recalled that when we did the evaluation back in 1993 it rained for thirty days straight. Pat swept Richard and me under her wing and into her chariot. Off to Chez Papillon!

Chez Papillon welcomed us to steam gently by a huge wood fire, and fed us *steak au poivre*—not the mere Parisian version, but brandied sauce with fresh green peppercorns of Madagascar. Of perhaps forty people in the restaurant, all were lily white, barring waiters.

That is, unless you count Prof. and Mrs. Toshisada Nishida. They were more greenish, having arrived two hours before from Japan. Direct via Singapore, Mauritius and Antananarivo to Fianarantsoa, airport after airport, just to preside over the Pre-Congress and Congress. The Papillon dinner for our small, select group was in their honor, but Toshi felt even more shattered on hearing that he would have to give a fifteen-minute welcome in English as president of the IPS to open the Congress, leading into a speech by the prime minister of Madagascar.

What really perturbed the Nishidas, though, was not airplanes, or the speech, or the hothouse of Chez Papillon.

It was that they went for a walk past a little market for their first glimpse of Madagascar. Elegant Mme Nishida spotted a butcher selling cubes of meat weighing 50 grams, and a vegetable stall doling out Chinese cabbage at a unit price per single leaf. The Nishidas thought they knew poor countries. Toshi has studied chimpanzees for thirty-five years in Tanzania—the Mahale Mountains National Park is founded around his work. But even in Tanzania, he protested, people buy meat by the half-kilo, and cabbage by the head. They were horrified to arrive in a country so poor that a family counts how many square-centimeter meat cubes they can afford in their stew, and one by one the cabbage leaves.

August 5, 1998.

Sun! Sun in Ranomafana! Patricia Wright the optimist wins again! The waterfall of the Namorona river hurls itself over the escarpment. A cascade of spray throws rainbows in front of the green rainforest. The uphill ricelands of Betsileo, and the downhill slash-and-burn of the Tanala, still have a precious belt of rainforest between, clinging to the vertical escarpment slopes.

The Park entrance is on an almost flat bit of ground. In the entrance hut there are Park T-shirts for sale, and a rainforest mural with creatures to spot, and a bulletin board announcing the ANGAP-imposed fees for tourists. Also a guide's schedule with lowest fees per day for guides in training up to a much bigger sum for specialist ornithological tours arranged by Loret. There were 8,000 visitors last year, and 8,000 already this year by August—doubling yearly.

So Loret is still here, chief guide! What about Émile, whose photo and name are not on the board, though you may still see him taking notes with Bernhard Meier in the August '87 issue of *National Geographic*? Émile is now head and mainstay of the Ecological Monitoring Team, guide no more, but a full-time researcher.

The porters assemble—median age about 12—and heroically lift our bags and knapsacks. They scamper down the main trail steps to the river—a few of the worst bits are now concrete. And there is the latest new bridge! In iron! High above most flood crests! The Namorona foams away beneath, pretty-pretty today, but not so after a cyclone.

Up the other side. We cut off on the trail labeled 'Research Personnel Only,' to not one but two log cabins. There is also a strange, isolated construction, containing two flush toilets, with a couple of oil drums of water outside and a bucket so one can carry out the flushing. What has come over Ranomafana?

The accommodations are tents, as before. More of them, numbered, each with a tarpaulin strung above to keep rain off the outer skin, and reassuring pre-dug ditches to sluice rain away from the groundsheet. Richard and I crawl into ours in a choice spot near the river to blow up mattresses and spread out sleeping bags.

Welcome lunch on the research house porch. As we shovel in rice and beans and mutton, Patricia introduces the students, about half-foreign, half-Malagasy. She tells us that more than 200 research papers have now been published on Ranomafana, and about fifty Malagasy have studied here, some going on for further work in the States. Undergrads from Stony Brook can spend a term of field study here. Earthwatch volunteers come to monitor sifaka, and this year, under Chia Tan, Pat's graduate student, to take on the much harder job of following all three species of bamboo lemurs.

Can we see? Oh, yes, says Pat. So all thirty of the Pre-Congress slippy-slidy up the trails still muddy from the previous week of rain, uphill (everything is uphill from the river), over a ridge, down—and there are the golden bamboo lemurs themselves, a male, a female and a six-months-old youngster.

They hang above us in the bamboo, snipping the cyanide laced tips and chewing them, just as they should do. Now the adults wear collars for research identification, but they have the same pudgy cheeks, the same round eyes and round bottoms, the glowing russet fur and golden faces, just as their forebears did when I first saw Ranomafana twelve years ago when people did not even know this was really a new species. Maybe that green-tagged male was the youngster I saw in '86—it is in the same stand of bamboo.

Professor Nishida stared and stared—two days before yesterday he was in Japan, and then entombed in limbo of air travel, and suddenly, as if dreaming, transported to a Malagasy rainforest with gold afternoon sunlight glinting on bamboo and filtered through all the layers of green, and among it three members of one of the world's rarest known species. The golden bamboo lemurs peered down placidly at the Nishidas and the thirty others from all the world. Ranomafana's lemurs have been followed now by watchers for all of their lives. They just eat, and scent-mark, and travel, as they always have, with brains that could never understand that without these human followers they would all be dead, their forest converted to manioc fields and the veneer of executive desks in Japan.

Supper with chatter and singing, and clinking glasses, then snuggling into the comfort of warm sleeping bags while insects and nightjars call in counterpoint to the rushing river.

August 6, 1998.

Madame Berthe Rakotosamimanana, general secretary of the Primatological Congress, drove down from Tana and arrived at 3 a.m. just to greet the Pre-Congress and tell about the extinct subfossil giant lemurs. Then she turned round to drive back to Tana—an immensely gracious gesture from someone currently trying to shoehorn 550 foreigners into a capital city with 400 hotel beds.

Pat's Malagasy students followed, with talks on forest transects and bamboo growth and flowering (it does not all flower and die, as in China, but dribbles along from season to season and always supports its lemurs) and a wonderful fruit guide with colored pictures of blue and rose and green globules. Of course fruit is *in*, now, since people including Pat have pointed out that birds, bats and primates are the nursery gardeners of the forest. Without them, no new trees.

In the afternoon a juicy demonstration by Ed Lewis, veterinary geneticist, on how to make, load and fire an immobilizing dart, including saying 'never point this at anyone' while waving the thing around in our faces. Some of the more sensible primatologists dived for cover. Richard said it confirmed everything he ever heard about vets. Young Seth Melnick, aged 10, scored a bullseye with the blowpipe—on a target.

Tucked up in sleeping bags; the rushing noise of the river gave way to a more insistent noise of bucketing water. Inevitably a small damp spot appeared just over our heads, where the shielding tarpaulin sagged onto outer skin of the tent under the weight of Ranomafana's rain.

August 7, 1998.

Wet boots, wet mud, wet bamboo to back into when going to the loo at night, damp sweatshirt from the drip in the tent. Richard murmurs, 'If you'd been working in Ranomafana, not Berenty, when we got engaged, I might have thought twice about asking.'

Patricia beams like the sun, and exhibits to all that *she* has picked up the first leech. Goody. Now, says Pat, we shall reschedule a bit. No lemurs today. Let's enjoy lectures! And then let's all go down to the town, where there is electricity, for lectures with slides!

The *pièce de résistance* was Sam Wasser's talk on what we can learn from poop. Hormones in feces tell him all about the reproductive cycles of Tanzanian baboons: who became pregnant and who aborted. He pointed out that there is a good season for weaning babies, but all the dominant baboons converge on that season. If a subordinate baboon gives birth at that point she is harassed so much that her infant is very likely to die. Subordinate baboons have infertile estrous cycles, or even early abortions, so timed that their infants are born at climatically worse seasons but socially safer ones. The bottom line is that some primates, quite likely including ourselves, know physiologically when there are social pressures not to reproduce. Sam pointed out that our doctors treat miscarriage as a medical and physiological crisis, a disease to cure, without asking what other social and mental stress may be involved. For some or many women, early miscarriage or infertility may be the body's judgment that reproduction just now will not succeed. Sam is, of course, working with clinicians who currently diagnose when fertility is best treated by simple physical means, medical brute force, and when other action to relieve women's stress may be even more important.

That was just the start. Then came reproduction in elephants, which involves spading through a couple of gallons of elephant splat to mix up the hormones thoroughly before you take your tube of sample to the lab. Then came population estimates: just count poops. Enter the sniffer dogs. The same dogs that detect an ounce of marijuana in your airplane luggage. Sniffer dogs can find wolf and grizzly bear poops under 7 feet of snow in Yellowstone Park in winter. Being dogs, they certainly find grizzly feces more interesting than mere contraband grass.

On the way home in the dark a pair of avahi peered down from above the mud-sodden trail like miniature koalas with searchlight eyes.

August 8, 1998.

I do not believe it. It is not raining. Our last day here, and it isn't raining. This is the day I pressed Pat to let us see a village. We will do so. So no more lemurs. Ho for development.

Except, at breakfast, a scout came to say that the *Hapalemur simus*, the greater bamboo lemurs, were practically in camp.

Six or eight of us rushed off uphill to find the *simus*. Rushed, in my case, means pulling myself along orang-utan-like by handholds on trees. I am getting not only so old but so heavy that I am positively embarrassed when young men have to hoist me upwards with shoves on the bottom. I rushed in spirit.

Patricia told us it had taken twelve years to habituate one group of greater bamboo lemurs. They had lurked unseen or flitted away as gray shadows. At last determined Chia Tan won their trust, or, rather more precious, their indifference. The group is currently just one male and three females. Amazingly the male is dominant over the females! A first for lemurs, if it extends through the species—but with only one group habituated, or indeed known for sure to exist, how do you tell?

As I came puffing up to the ridge top, all the primatologists sat in a semi-circle round the male. He sat at the base of a 6-inch bamboo stem, splintering loose shreds like a bouquet of dry spaghetti. The male put his head sideways on, his lips drawn up in a snarl, and attacked the stem with canines and the lower, caniniform, premolars. He tensed his neck muscles and ripped loose another section. Primates just don't act like that. Beavers do, but beavers have the right tools. It's like cutting trees with opposed Swiss Army penknives instead of a saw.

Occasionally he glanced up at the battery of primate watchers inching up to a mere 2 meters away with a twenty-one-gun salute of camera flashes. Habituated, in this case, means blasé to the point of insult.

My head whirling, I slid back downhill to do my duty and talk to villagers. But the *simus* was worth the whole trip.

Later.

Uphill, not downhill. Uphill we rode—the naughty lemur-watchers who missed the bus benefited by our own Land Cruiser. So we would see Betsileo rice-growers, not disgruntled Tanala like the people of the Village of the Fig Tree. We leveled off and sailed right past the village of Vohiparara at the head of the escarpment. That was the village where the evaluation team of 1993 shivered through an hour or two of lectures dedicating the just-completed school, the day after a frost had burnt all the banana trees brown, and we huddled cowardly in macs and welly boots while the village president sat upright and immobile with his bare gnarled feet on the not-quite-dry cement floor. I hope they have got rid of those earth-worm-sized intestinal nematodes.

We carried on another half-hour of potholes, to Sahavondrona. A farmer was showing the gang how to plant rice.

And that is new.

Back last year, Professor Norman Uphoff of Cornell University collared Richard at a meeting and said, 'Let me tell you about planting rice! It is spectacular!' Cornell took over from North Carolina State as the development side of the Park. Just as predicted, there was a two-year gap for further studies and a new departure, rather than follow the lines of their predecessors. No more talk of cat latrines.

Norman Uphoff crowed, '8 or 10 tons a hectare where they were getting 2 before! No, no fertilizer. It is all in a method devised about twenty years ago by a Catholic priest in the Ranomafana region. About a thousand farmers have taken up the method this year, and it is spreading like wildfire. The most exciting part is that this is new, brand new, done nowhere else. The discoveries we are pioneering in Ranomafana could revolutionize world rice culture!'[1]

The key is early transplantation. Traditionally, Malagasy transplant rice seedlings from the nursery fields when they are about 20–25 days old. This is just the age when the seedlings put out new tillers and grow the seed-bearing stalks called panicles. Transplanting is traumatic for any plant, of course. It sharply discourages them from making panicles. Instead, if you move them as young as 5 days old they have time to settle into their new home, and then they panicle like crazy. So more seed heads, and much more grain.

To make it work you must treat the tiny sprouts tenderly. First, space them well apart, both within and between rows. The Betsileo farmer boasted, 'Before, we needed 15 kilos of seed per *are* (100th of a hectare, 0.025 acre). Now we need just 5 kilos!' That matters a lot to a subsistence farmer who could eat or sell the other 10 kilos.

Next, you mustn't drown them: lower the water levels in the paddy fields. And, finally, you must weed thoroughly three or four times a season. Deep water used to keep down the weeds, but with lower water you must do the work

1. Father Henri de Laulanié de Saint Croix, agronomist, university teacher at the Angers School of Agriculture, and Jesuit priest. SRI rice: http://sri.ciifad.cornell.edu.

yourself. The downsides are obvious: more work both at planting time and at weeding. Also, though no one has mentioned this to me, higher risk. Transplanting tiny vulnerable shoots must lay you open to losing the whole crop in one cloudburst or five days of drought. High-risk crops, in turn, reflect the growth of social support structures through the government, or through aid from, say, the National Park, or just through families spread out into different regions and sectors of the economy. With a social support network it is well worth shooting for 10 tons per hectare.

Next, the village president showed us the granary. It looked like a small cement house with no windows and a padlock. Cornell made a microcredit grant of a couple of hundred dollars for cement to build it. Villagers bring bags of grain to store, each labeled with their name. Before, they had to sell the grain after harvest at the year's lowest prices. They wound up buying back food in the *soudure*, the evil time between planting and harvest, when prices are highest. Now they can save their grain and sell it themselves during the *soudure*. About a third of the villages round the Park have now received such microcredits, derived from Park entrance fees, and mostly used for granaries or tiny irrigation dams.

Education is a real worry for the village, because for several years there were *dahalo*, bandits, cattle rustlers, along the 5 km walk to the nearest school. There is a five- to eight-year gap: the current teenagers never learned to read. Now they have an elementary school of their own, and other villages' children walk to them. I asked what the *dahalo* were after. Zebu cattle. Most of the families used to own twenty zebu, which they pastured in the forest. Now, only a couple of families are left holding three or four. (I wonder if this is one reason this village was so eager to embrace the benefits of the Park?) Someone else asked the more telling question: did they know the identities of the *dahalo*?

Of course. They were men from the village up the road. Some had been arrested since, but not all.

We forget so easily how easily law and order evaporates.

Still later.

A fine speech by Jocelyn Rakotomalala, the new, enterprising and apparently honest national director of the Park. Rakotomalala admitted there are bound to be problems: a third of the villages have profited from the Park but two-thirds still have not, so the jealousies must be faced. He sees this as his major task as the new Park director.

We trooped over to the Hotel Thermale, the much decayed relic of Ranomafana's former life as a French spa. We sat on chairs in the main room while Richard unveiled his flip chart of the decline in Madagascar's GNP—40 percent since the high of 1971. In depth and in duration that makes the Great Depression of the thirties look like a hiccup. Mad. is now the eighth poorest country in the world, per capita. Richard was blistering about demands that a country in this state pay its international debts to the bankers, the Bank and the IMF. Nations are not allowed to go bankrupt, which means that they can be squeezed forever.

Lectures gave way to dinner with the *prefét* and the *sous-prefét* and the mayor and the army. Patricia is in her element, toasting the people she has worked with to bring her park into being, and actually in tears of joy that the Pre-Congress has brought the world to Ranomafana. President Nishida replied gracefully, although still wonderstruck that he is actually in Ranomafana and not Kyoto.

Dancing! A local dance troupe of men, women and little tiny boys. Their dances were a weird parody of military drill, carrying 'rifles' hand-carved of wood. Again, the memory of '47 with the French military conducting search-and-destroy raids out of Ranomafana. Strange that this is the piece of their history enshrined in local dances.

We all rolled over to the thermal swimming pool for swimming by moonlight in the sulfurous waters out of the ground. The hot water comes straight out of the forgotten geological era when the earth's fires threw up the escarpment instead of merely warming a pool full of splashing foreigners led by the Nishidas—at last allowed a proper Japanese bath. Lovely. But while the Pre-Congress swam, I lay on a concrete bench and slept, eyes closed but still seeing the full moon sail in and out of clouds that dropped a gentle rain on Ranomafana.

August 10, 1998.

Hilton Hotel, Antananarivo. Madame Berthe was dancing. The silver threads in her black evening dress twinkled like the silver threads in her upswept hair. Long ropes of pearls and gold swung on her ample bosom. Her high heels tapped and jiggled.

About five hundred of the world's primatologists gyrated on the Hilton ballroom floor, holding hands, approaching each other in a scrimmage of opposed conga lines, snaking single file with hands on hips under a London Bridge line of couples clapping hands above the snake-line, twisting and rocking in singles and couples and trios and impromptu circles. Show-offs gravitated to the center of the circles: minute Wesley from southern India, bent backward in a bow with his hands clutching the air; Sam Wasser the poop analyst from Oregon, his black beard bristling and wagging in counterpoint to his hips; a student conference volunteer from Antananarivo, her waist-length hair shimmering in time to her seductive plateau Malgache shoulder-shake.

Madame Berthe Rakotosamimananana had brought them all there. For her, the whole of the Hilton danced in triumph.

We had reached the final evening of the International Primatological Congress of Antananarivo. Nobody but Madame Berthe really believed it could be done, but she did it. She and the GERP, *Groupe d'étude et de recherche sur les primates de Madagascar* (the Madagascar Primatological Society). All nine members of it. They invited what turned out to total 550 scientists to a city with 400 hotel beds, broken-down buses, and a university in advanced decay—and got away with it.

I did in fact know the size of the GERP when I was president of the International Primatological Society. I urged the IPS Council to accept Madame Berthe's bid to host the next Congress. It did not seem necessary to raise doubts. The rest of Malgachisant primatologists did not unduly point it out, either. We thought the Congress would fly in spite of meeting in the eighth poorest country in the world because Madagascar itself is so beautiful and everyone could go away and watch lemurs. We did not expect to close the meeting whirling around the Hilton floor, hugging and grinning, me and Richard and Hilary Box in her rugger-striped jersey and Kaberi Kar Gupta, all four and a half feet of her resplendent in her gold and orange sari; Leticia Dominguez Brandão doing Brazilian sambas; and Hanta Rasamimanana, the program chairman, mother of three teenagers, shimmying with Canadian Lisa Gould, who turned out to have studied belly dancing. Mukesh Chalise towered above them with a beard to outmatch Sam Wasser's. Mukesh had just received the Society's conservation award for mobilizing the world online, persuading the King of Nepal to forestall the slaughter of 500 of the world's rarer monkeys. His grin lit up the dance floor.

What did Berthe and the GERP do to make it happen? Fix the University to start with. A bit like trying to play the *Moonlight Sonata* by first building the piano. Last November when I was leaving Mad. she asked me to lunch at her family's on the way out, saying, 'I'm all right; it's Wednesday.' 'Wednesday?' 'I have had yesterday to recover.'

On Thursday the week before, the Office of the Budget called up and said, 'Mme Berthe? Where are your estimates?'

'Estimates of what?'

'The documents for the PIP, this year's budget for capital investment.'

'But what documents?'

' You should have already given us your estimates for the refurbishment of the University lecture rooms.'

'Wait-—there is some mistake. I am a professor of paleontology.'

'Are you Mme. Rakotosamimanana, Berthe, *secrétaire générale* of the International Congress of Primatology?'

'Yes, I am.'

'In that case kindly hand in your estimates on the proper forms, in quintuplicate, by 5 p.m. on Monday when the year's PIP will be finalized.'

Berthe is not quite an ordinary professor of paleontology. Back when she was permanent secretary of higher education, and even in the days when she was forced to refuse foreigners' visas as the Tiger of Tananarive, she could get things done. She whistled up a consultant expert in the filling out of government forms, pulled out the list she just happened to have already made of the needed repairs, worked 48 hours a day all weekend, and sauntered into the Ministry at 5 minutes to 5 on Monday to slap the forms on the desk.

So when the primatologists turned up the rooms were painted and light fixtures no longer lethal and toilets newly installed and equipped with paper. (I doubt there was a working toilet in the place last year. People used darker spots in the corridors, like Louis XIV's Versailles.) All the plumbing is new, to last in perpetuity or until the next Mme. Berthe overhauls the University again.

When we arrived my colleague Hanta had the scientific program printed, and the abstract booklets ready, and a pile of ultra-chic raffia and leather briefcases to hold everything. The caterers laid on a 20-meter buffet of luncheon, and the projectors in every room almost worked.

And there were so many good talks! Too many to tell...

For the opening ceremony Madagascar's prime minister made a great pro-environment declaration, and President Nishida of the IPS gracefully explained what the Society is, and Joseph Andriamampianina, now retired as head of the National Office of the Environment, orated about how emotional he felt that we should be meeting here. I remember ending the *National Geographic* article on Mad. in 1987 with him with tears of joy in his eyes at the opening of Bezà Mahafaly Reserve, the first new reserve in twenty years. Well, why not—if what has happened in Madagascar isn't worth weeping and whooping about, what is?

Closing Day: Elwyn Simons, Prithajit Singh, Yves Rumpler, Jean-Jacques Petter and I were hauled up to sit on stage, while Madame Berthe chaired the meeting. The ministers of agriculture and the environment spoke their pieces. I had thought it would be pure formality. Then it seemed there would be something more... Mme. Berthe asked me to stand up and come over—the very small doyen of the University muttered 'By virtue of the authority vested in me by the President of the Republic of Madagascar, I name you Officer of the National Order of Madagascar!' And he pinned a great big medal with a green and red ribbon and a superimposed rosette onto my red coat, almost as big as first prize at a pony show. Ten or twenty flashes went off, so it seems that everyone at the Congress has a picture of me with my jaw falling open in surprise like a hooked pike.

Next, Elwyn, Prithajit and Yves were made Chevaliers of the Order. We could see there were only four medals pinned onto the cushion to give out, and there was Jean-Jacques sitting next to me all undecorated. I whispered to him, 'You have one of these already?' He whispered, 'I seem to think so,' but still there he was…

Madame Berthe stepped to the podium. She began to read Jean-Jacques's list of services to Madagascar, including teaching at the University for a couple of years and organizing the very first International Conference on the Environment in 1970. So, said Mme Berthe, 'we now officially name you a professor emeritus of the University.'

In my world a professor emeritus is not so grand as a knight of the legion. However, the tiny doyen of the University was wearing an academic robe the like of which I've never seen. The usual flowing black, but a row of little black buttons down the front, cerise satin lining, bell sleeves with deep cerise facings buttoned to the black in flounces, a white pleated jabot at the throat, and over the left shoulder a token strip of *lamba* of the richest deep-woven cream-colored silk, held to the shoulder pad by a crimson seal the size of a butter-plate.

Furthermore, this apparition had another such garment over his arm, which he placed upon Jean-Jacques, though slightly impeded by the fact that Jean-Jacques is 6 feet tall. Clearly designed in the days when few Malagasy reached 5 feet. (Richard tells me statistics on stunting and the growth of stature as a mark of the world's emergence from poverty, and I note that nowadays I do not automatically see over the heads of everyone in a Malagasy crowd. Not even professors emeritus.)

Jean-Jacques strode out of the hall trailed by all us new-made knights, with his shoulders constricted back like a straight jacket with handcuffs and the gown swinging way off the ground as a bum-teaser.

So we all went off to the Hilton, to dinner and dancing and dancing and dancing, even Richard and I suddenly able to swing with the joyous crowd, and Madame Berthe dancing happiest of all.

Alison with her parents Alison Mason Kingsbury
and Morris Bishop, circa 1948

Alison with toy monkey, aged 4, and posing for her mother, aged 6

Alison with sifaka, Yale, circa 1961

Alison with kinkajoo and brown lemur, Yale, circa 1961

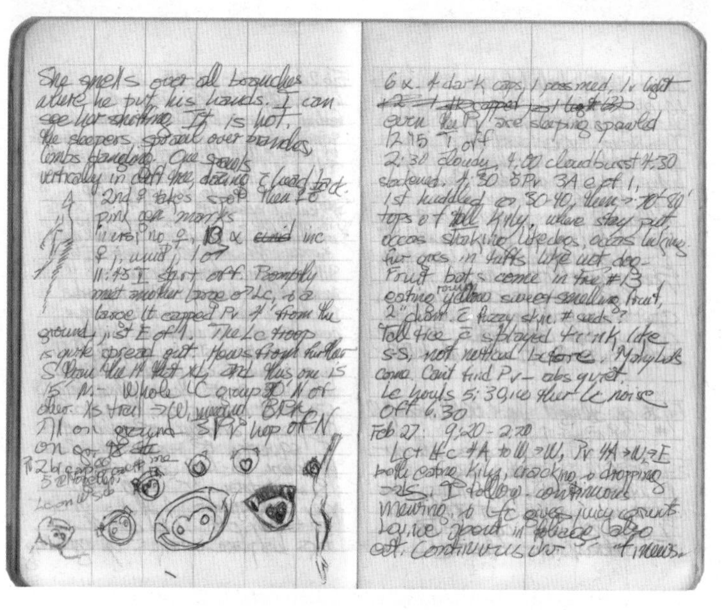

Alison's early Berenty notebook, circa 1962

Alison on the cover of her first book, *Primate Behavior*, 1966

Richard and Alison Jolly on their wedding day, 1963

Alison Jolly in her trademark sunglasses, 1983

Alison with her children, circa 1972

Outside the family home with best friend Jiffy, 1978

Activists for Unicef: with husband Richard, 1983

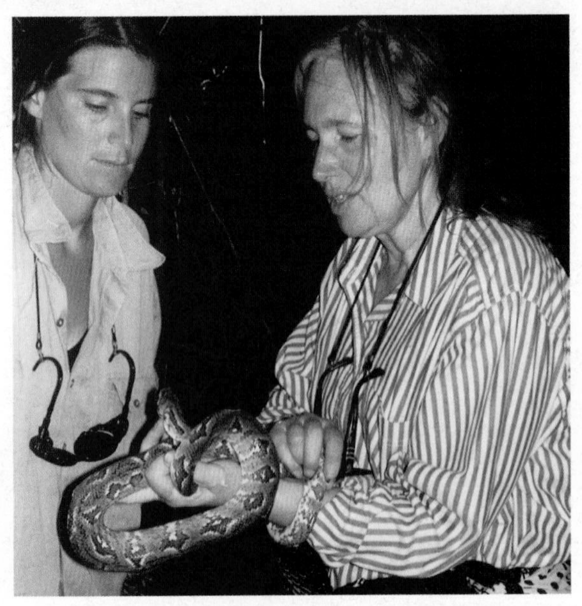

All life is marvellous: with Helen Crowley,
manager at Berenty Reserve, circa 1992

Alison and Hanta with Earthwatchers at the entrance to the Berenty Reserve, 1984 (Hanta is fifth from left, Alison eighth from left, back row)

Alison laughing with friends at Fort Dauphin, 1987

The Duke of Edinburgh, as WWF president, helped launch modern conservation in Madagascar in 1985 (left to right: Joseph Randrianasolo, minister of Water and Forests; Russ Mittermeier of Conservation International; the Duke of Edinburgh; Berenty Reserve owner Jean de Heaulme; ornithologist Mark Pidgeon)

1987 ministerial visit to the USA to tour American zoos. Includes Joseph Randrianasolo, minister of Water and Forests, with Joelina Ratsirarson; Henri Rasolondraibe, secretary general of the Ministry of Scientific Research; Voara Randrianasolo, director of Tsimbazaza Zoo; Barthélémy Vaohita of WWF; and Berthe Rakotosamimanana of Higher Education—with her ever-present handbag

Joseph Andriamampianina, Alison Richard, Pothin Rakotomanga,
Bob Sussman and forest guard joyfully celebrate the
designation of Bezà Mahafaly as a Special Reserve, 1985

Patricia Wright and Loret Rasabo, Ranomafana, circa 1988

Russ Mittermeier and *Hapalemur*, Ranomafana 1986

Josef Bedo and *Argema mittrei* luna moth, circa 1989

Alison with daughters Margaretta and Susan at the opening of the Alison Jolly lemur sanctuary in Apenheul Primate Park, Netherlands, early 1990s

Alison and Richard with son Richard, mid-1990s

Alison with son Arthur in Madagascar, early 2000s

Alison climbing onto *The African Queen*, Masoala Peninsula, 1985

Alison with men working in sisal plantation, Berenty, early 2000s

Alison observing ring-tailed lemur, 2005

David Attenborough with Melanie Dammhahn,
Kirindy 2010

Alison with David Attenborough, Kirindy 2010

Hanta and Alison in a boat on Lake Alaotra

Alison receiving honorary doctorate,
University of Antananarivo 2012, with Hanta

New generation of Malagasy biologists and conservationists, including Alison's students Josia Razafindramanana (second from right) and Hajarimanitra Rambeloarivony (far left), circa 2010

Alison in heaven, circa 2005. Though she and others have instructed no feeding of bananas to lemurs at Berenty, occasionally they steal some

Rainforest, Ranomafana National Park

Berenty Reserve at night

Deforested hills, central Madagascar

Zebu cattle drove

Looking out from NamanaBe Hall, Ranomafana National Park

Baobab tree

Waterfall in Ranomafana National Park

Antananarivo, with Queen's palace on hilltop

Sportive lemur, *Lepilemur mustelinus*, in tree hole, Berenty Reserve

Logger felling tree in rainforest,
Bay of Antongil, northern Madagascar

Man cutting baobab bark for rope-making,
Andavadaoaka, western Madagascar

Antandroy leading zebu cattle to market in Amboasary

Rekanoky, chief reserve guardian at Berenty, cradles a two-gallon egg of the extinct elephant bird. Behind him are 'octopus trees' of the spiny forest

Daubentonia madagascariensis, more commonly known as the aye-aye, showing its third finger, specialized for insect-grubbing

Sportive lemur—the name 'sportive' came about owing to this species' habit of adopting a boxer-like stance when threatened

Verreaux's sifakas in trees, Berenty Reserve

Verreaux's sifaka leaping, Berenty Reserve

Hapalemur aureus, more commonly known as the golden bamboo lemur, eating cyanide-filled stalks, Ranomafana National Park

Indri, one of the largest living lemurs in Madagascar, singing at Andasibe

Mother and baby ring-tailed lemur, *Lemur catta*, Berenty Reserve

Mother and baby ring-tailed lemurs playing in trees

Geneviève Sailambo holding a baby ring-tail

Advance of a troop of ring-tails, Berenty Reserve

Two ring-tailed lemurs

Madame Berthe's mouse lemur

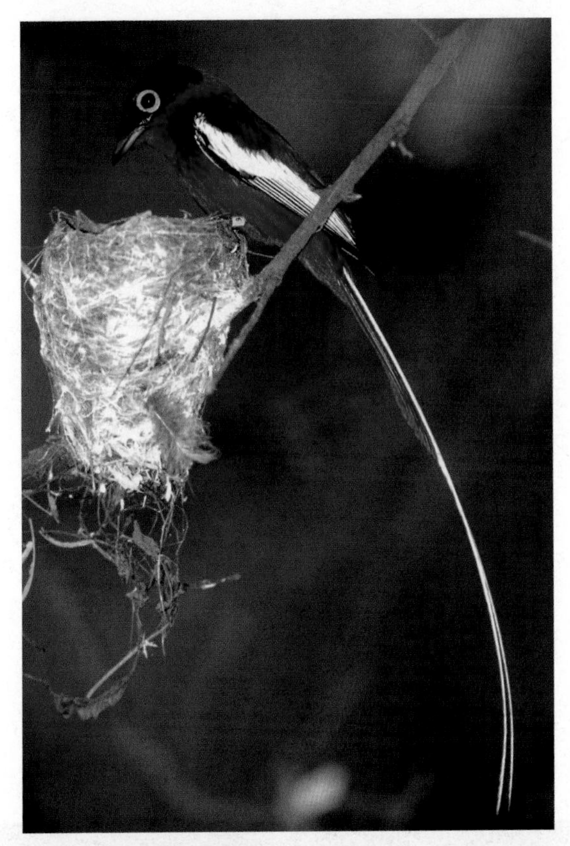

Gobe mouche bird of paradise

Fosa eating a pigeon

Parson's chameleon male, *Calumma parsonii*, eastern Madagascar

The endangered radiated tortoise, *Astrochelys radiata*

PART IV

Weather

Madagascar has one of the most unpredictable of climates: cyclones, droughts, floods. Traditionally both wildlife and people have confronted this unpredictability. This isn't just farmers' chronic grumbling about the weather. Malagasy actually have greater reason to complain.[1]

We now move away from rainforest to the semi-arid spiny forest of the south. This also brings us up to the present and future time.

People struggle to describe that forest: Surreal! Lunar! Undersea corals growing in a desert! It is dominated by 20–30 foot tall *Didiereaceae*, members of a plant family unique to Madagascar—cactus-like spires pointing to the sky, but in no way cacti. They have wood inside, not spongy pulp: wood that builds the tiny plank houses of the people who live there. Each plant is adapted to store water and avoid the worst of the sun's rays. You push through spike-studded thickets in a forest of no shade. Plant ecologist Peter Grubb claims there is no place in the world that looks like the spiny forest of Madagascar, except perhaps one valley in Mexico which has

1. R.E. Dewar and A.F. Richard, 'Evolution in the hypervariable environment of Madagascar' (2007).

a similar climate of wet and extremely dry seasons with sporadic showers to maintain life through the dry months.[2]

Spiny forest is arguably the oldest ecosystem of Madagascar. It probably originated in the Cretaceous period after Madagascar split from Africa and before the extinction of the dinosaurs, when Madagascar lay in cold, arid latitudes. Rainforest came later as the island moved north toward the warmer, wetter tropics. Nowadays Madagascar straddles the 'intertropical convergence' where cyclones are born. The remaining spiny forest is confined to the region around and below the tropic of Capricorn in the rain-shadow of the eastern mountains. It has the highest endemism of any region, even for Madagascar: some 95 percent of its plant and animal species exist only in the South.[3]

Combine a semi-desert climate with climate unpredictability: you have a recipe for killing famines that alternate with years of recovery. Chapter 19 tells of the dreadful famine of 1991–92; also of the historical record of earlier crises, and the change that has come as people of the South learn that they may call on international food aid... over and over again.

Chapter 20 gives the strategies of ring-tailed lemurs and leaping white sifaka confronting a climate where any one lemur lifetime is likely to include a catastrophic year. There is data from Berenty, the reserve where I work, and from Bezà Mahafaly Reserve in the western part of the south. Chapter 21 tells of the human social system of Berenty in a plantation economy dating from colonial days. In contrast, at Bezà Mahafaly scientists have worked together with villagers for the past thirty-seven years to improve welfare for all. It asks if these tiny reserves can be extended to the vast reaches

2. P.J. Grubb, 'Interpreting some outstanding features of the flora and vegetation of Madagascar' (2003).

3. H. Perrier de la Bathie, *La Végétation Malgache* (1921); N.A. Wells, 'Some hypotheses on the Mesozoic and Cenozoic paleoenvironmental history of Madagascar' (2003).

of Tsimanampetsotsa, the 'Lake with No Whales'—and concludes that large-scale efforts face all the same problems that Ranomafana did in spite of very different geology, weather and landscapes. Chapter 22 finally confronts the long-term question for both the dry south and for the rest of Madagascar: if life for both people and forests is unsustainable now, what will it be in the future?

NINETEEN

Famine in the south

There have always been famines in Androy, the Land of the Thorns, sometimes called the Land of Thirst. When the rains fail, crops fail. People sell their hoarded cattle and search for edible roots in the forest. They slake the acid pods of tamarind trees with chalk or wood-ash to make famine food that gives bellyache but staves off death. Children die first, then old women, then old men. The able-bodied emigrate to any town with a hope of food. Men may find work, but at least wives and daughters can scrape a pittance as prostitutes. If one is about to die utterly destitute, turn Christian. Christians can be buried without the traditional sacrifice of zebu.[4]

People have always known it would happen. The list of major famines goes back and back. 1991–92, one of the worst double El Niño years; 1983–84; 1970; a gap in the records I don't know how to fill in the 1950s and 1960s; 1943–44, again a double El Niño year; 1937; 1930; 1921; 1913; 1911. There have been many El Niño years without famines, but rarely a famine without El Niño. The peak El Niño of 1997 brought low rain, but that was only for a year, so people tided over.[5]

---

4. R. Ramahatra and H. Patterson, *Poverty in Madagascar* (1993).
5. S. Frère, *Panorama de l'Androy* (1958).

After each series of bad years, though, the rains return and the people flourish. Tandroy raised in years of plenty were traditionally among the tallest and strongest in Madagascar.

In 1991–92, that changed. The international world noticed there was famine in the south, almost for the first time.[6]

September 19, 1991. Berenty Reserve.

Photo. 4.30 p.m. The sun dead center, low, like an orange eye. The sweep of the wide flat river bed, with a curve of dark forest on either bank. The bank with Berenty Reserve is rich with birds, lemurs, undergrowth. The far shore is a canopy of trees where zebu graze on packed earth below. In the photo you can't see the difference: just two dark crescents converging as they curve round a quarter-mile width of sand.

The river bed is dry. It is like a piece of the Sahel a quarter of a mile wide, the Sahara where a river should flow. Its yellow surface is pocked with the hoof prints of zebu led to drink at the one last stagnant puddle in the center; also with the pits where people have dug for their own water.

Walking up the river bed, toward the sun, the silhouettes of two tiny figures: a man and a woman walking. The woman bears a 3-foot pile of their goods on her head.

Oct. 4.

Photo: A group of eight women with buckets and gourds on their heads trudge from the pit in the center of the river their men have dug.

And photo group: The women climb the steep river bank. Our group of camera-hung whites asked if they would mind their pictures taken, for a present. *So click* the row of eight

6. Not quite true: Raymond Decary, the local administrator of Androy, persuaded the colonial government to aid the destitute in the great famine of 1930. However, 1992–92 was the first international response. A. Jolly, *Lords and Lemurs: Mad Scientists, Kings with Spears, and the Survival of Diversity in Madagascar* (2004).

standing stiff and formal in African picture pose, each with a bucket or gourd on her head, topped with a tuft of green leaves to keep it from sloshing.

Close-up: A girl of perhaps 15, probably with only one or two children so far, smooth chestnut-colored cheeks, hair in elaborately bunched braids behind each ear, a round brown calabash on her head, the same color as the dirt-dun rags she wears as a dress.

Close-up: The end woman, short, almost a dwarf, in a loud pink dress torn on both shoulders and a bright blue water bucket. She grins hugely and strikes a pose, hands on hips and one foot forward. The oldest receives 5,000 francs (about $2.00) to share among all. They seem delighted, and turn to the 5 kilometer trek home.

Oct. 1.

John Davidson, USAID, Economic Attaché 'We've known since April there could be famine in the south. Médicins Sans Frontières gave the alert back then. So we have acted long ago. A shipment of 15,000 tons of rice is due in Tulear harbor in the second week of October.

'This rice shipment and famine relief is a major, multi-donor operation. You should talk to David Fletcher of PAM, the World Food Programme.'

Oct. 9.

David Fletcher, World Food Programme 'This is a very large project. WFP is coordinating, but there are eight donors in all. USAID is the major one, with its 15,000 tons of rice, but there are Germans, Swiss, UNDP… Whatever you said to the officials at Unicef seems to have fired them up, Alison, and they're coming in on the transport costs. It isn't very sexy for Unicef to do transport, but that is what we need.

'I have hopes of the World Bank coming in, too, but José

Bronfmann (the World Bank rep) is not at all convinced. In fact he read me the riot act about food aid skewing local markets. He'd only be reconciled if it were food-for-work. I'm all for food-for-work projects, but the south has been so neglected in development, and the crisis is so imminent that we can't hold off to identify and set up work schemes unless there are obvious ones to hand. I happen not to share Bank philosophy.

'I should add that the local markets are skewed anyway. This year 23,000 tons of maize and 9,000 tons of manioc were exported from the port of Tulear. A colleague discovered in February that 1,500 tons of manioc were waiting in Tulear for export to France—for animal feed. We tried to get an EC loan to buy it back for people, but it took so long that the manioc was exported anyway. What we did get was a gift to buy 570 tons of maize in the south for urgent needs, which will help keep the price up.

'How well do you know Jean de Heaulme?'

Me 'Very well.'

David 'What is he like? I have never met him.'

Me 'Sweet, courteous; no, courtly, devoted to the south... Why?'

David 'The reason I wanted privacy is that I just learned yesterday there is a problem with the 15,000 tons of rice. You can imagine that that quantity is coming on a very big boat indeed. Well, it's too big to get into Tulear harbor.

'We've calculated that if we can offload 4,500 tons it will lighten the ship enough to fit past the reef at Tulear. So it seems the first port of call will be Fort Dauphin—this week.'

Me 'Will it fit in Fort Dauphin?'

David 'They load the sisal with lighters. They can unload rice the same way, however far out it needs to anchor. But we need real mobilization, and de Heaulme is obviously influential.'

Me 'OK. If you want my opinion, he is a prince among men. Of course he's a businessman, and I don't know what he needs to do to operate in this country. But I do know that he has an extraordinary commitment to the South—people and land. One part of founding their sisal plantation was the de Heaulme family consciousness that they were providing a buffer against famine in the Mandrare Valley. There have always been famines—one particularly terrible one in 1943–44 when the Brits were blockading the ports of Vichy French Madagascar, and no food came to the south. Monsieur de Heaulme told me several times about being out as a boy with his father and seeing something waving from a tree. When they came up, it was a dead woman's hair. She'd climbed up into the tree to die so her body wouldn't be eaten by the dogs, which had gone feral from the villages because they were starving, too.

'The de Heaulmes are really more feudal lords than capitalists, with the better side of feudal obligation to others.'

David. 'We're in a real bind. The port warehouses at Fort Dauphin have the last World Food Programme food rotting in them now. I need to jump those warehouses and move this rice quickly—and intact—so you have helped...

'And just before I go, please give my regards to your husband. He won't remember me, but we met at the first UNA Conference on Disarmament and Development. He was the main speaker, and I was just out of university. It was his lecture and his advice that really started me on my career in developing countries.'

SOS: Save Our South!

David Fletcher went on national news with a video he took with his own camcorder: skeletal people and blasted crops. A photo appeared on the only newspaper's front page: a wasted teenager, with the caption 'Is this Ethiopia? Somalia? No, he is in Madagascar!' Plateau people dug into their own slim pockets to fund a national organization with an English name: Save Our South, SOS. Malagasy medical students and nurses banded together as ASOS, an NGO for emergency relief. Even a group of Antananarivo street children came to offer a few tiny francs as their contribution. The people of the capital did care, once they knew what the south was suffering.[7]

Oddly, when the head office of WFP in Rome heard about this, they were horrified. WFP had no authority to raise money within a suffering country, only to accept gifts from donors. This rule has since been changed. But by the time Rome found out what was going on, Fletcher was far ahead of them.

He spoke to the Swiss ambassador. Switzerland promptly sent two logistics experts from the Swiss Disaster Relief Unit, seconded for an entire year. They worked out how the trucks of Fort Dauphin could switch over from carrying sisal to sacks of rice and manioc for distant villages, mostly straight from the port, bypassing the warehouses with their temptation to steal. That helped people to stay at home and preserve their social order, rather than trek to handout centers in the towns.

The crucial link with the villages was an extraordinary team of twelve Malagasy, headed by a medical doctor, Arthuro Razanakolona Randrianiaina. They went to villages ahead of the food trucks to explain what would happen. In those days, international emergency food was typically unloaded by underpaid laborers with no great interest in the outcome—who might not even show up, leaving the cargo stranded. Dr.

7. Much of this drought chapter is from converstions with David Fletcher, pers. comm., 2013. ASOS: Aid to Save our South, continuing as Action Socio-sanitaire et Organization Secours. Swiss Disaster Relief Unit: now the Swiss Humanitarian Aid Unit.

Arthuro's team announced that if volunteer village labor did not appear within twenty minutes of a food truck's arrival, the truck would turn around and leave—rather spectacularly motivating.

Another key to the program was the de Heaulme trucks and drivers. Mobilization needs something to mobilize, and the de Heaulme empire was the only organization capable of responding. Jean de Heaulme takes great pride that over 95 percent of the donated food actually reached its destination. In most famines, aid agencies are proud if half of it is delivered as promised to remote rural areas.

Japanese Aid built a huge underground water cistern by the Mandrare river. The Japanese and Unicef provided cistern trucks to provision thirsty towns and villages. A host of foreign NGOs like Aide Contre la Faim arrived, with earnest European volunteers to run shelters and feeding programs.

Still the crisis deepened. The rains failed for a second year. People began to call it the drought of the century. In 1992 I went with Unicef far from Berenty, to village food distribution, to hospices for mothers with seal-eyed toddlers in every stage of malnutrition, to a village which once owned 500 zebu and now had 20 (prices of cattle on the hoof plummet in famine time, even while prices of any kind of food skyrocket). Most of the time I spent at Berenty Reserve with the Earthwatcher volunteers, eating lobster and roast beef, with cheese omelets for the vegetarians. As economist Amartya Sen brilliantly trumpets to the world, there is always food for those who pay. Famines are not about drought, but about entitlement. Even in Androy people did not die because there was no food available. They died because they could not afford to buy it.[8]

In 1992 half a dozen doctors and nurses happened to join one Earthwatch team. Earthwatch volunteers came to help

8. A. Sen, *Poverty and Famines: An Essay on Entitlement and Deprivation* (1981).

with lemur research, unaware of the famine. We all lived in the tourist quarters at Berenty cocooned from the world outside the forest. Then we went to see the hospital of Amboasary. The doctor in charge of the hospital greeted us. He showed us around filthy wards, the glass long since gone from the windows. He picked up a 2-year-old in his arms, unconscious, flaccid as a tiny rag doll. 'We will save this child,' said the doctor.

> I want to show him to you because he has just been brought in. This is starvation. But even in this condition we can save him, with feeding little by little. What we cannot do is change the surrounding state of the countryside, or the state of Amboasary town. I was trained in France: I have an education like yours. It has taken me two years to realize that my one medical goal in this town is to persuade the people to use latrines. Nothing of the fancy medicine I learned, the medicine I would like to practice. Latrines. If we succeed, we may, we may just, avoid the epidemics that follow famine.

It was that same day that we saw a family walk in totally naked to the soup kitchen run by the Sisters of Charity, who were cooking maize porridge in empty oil drums. Three of the Earthwatch medical people who had never been in a developing country before were too shocked to even speak for the rest of the day. Jim Sellers, another Earthwatcher, seemed strangely confident striding among the temporary huts built of straw in the famine camp. He went away and helped set up an NGO to provide medical volunteers and supplies for Madagascar.[9]

The overseas famine volunteers congregated in Fort Dauphin every weekend. All week they spooned gruel into dying babies, comforted emaciated mothers, fought the

9. www.tascmadagascar.org.

breakdown of recalcitrant food trucks, checked tallies to guard the food from going astray. Many were in culture shock to learn that when village families distributed food, it went first to men, then to women, then to healthy children, not to the smallest or the sickest ones. This is power relations, of course, but, in harsh Darwinian calculus, breeding adults are the most valuable survivors. Rephrased in cultural terms, that means the clan must live. Volunteers who had crossed continents to save suffering children dealt all week long with the knowledge that in some ways the clan's cattle matter even more. On Saturdays, the aid gang danced all night long at the Panorama Disco of Fort Dauphin to recover...

People who are far from home and under stress can abandon every rule of their own culture. The aid gang swapped partners in a kind of frenzy. Everybody knew everything about everyone else's affairs, to boot. The Antananarivo plane on Sunday brought in fresh potential mates, and swept others away to the sound of jealous screams and breaking hearts. Not to mention hair-pulling, eye-gouging fights outside the Panorama Disco. Social life in Fort Dauphin has never been so *interesting.*[10]

Finally, in October of 1992 the rains returned. People went back to their villages to plant crops, though many needed a gift of hoes even to break the soil—they had sold everything to survive.

The World Food Programme then had an intense internal debate over whether to pull out when the crisis ended. Were they just mandated to do emergency relief, or to take on long-term development? In the teeth of opposition from headquarters and from the Food and Agricultural Organization, Fletcher convinced WFP that Androy is a region which must recover from one famine in order to confront the next one. He

10. Helen Crowley, Reserve manager, interview 1999.

saw both a moral and a practical obligation to continue WFP aid as food-for-work projects, chosen from proposals made by villagers themselves for initiatives like local irrigation or co-operatives to market crops. That might encourage an increase in the entire food supply of the region. Again he had to fight distant bureaucrats who tried to mandate that Dr. Arthuro's team, with all their village contacts, were emergency workers who could not turn to becoming development experts. But the program went ahead. Unicef thought the same; it set about drilling boreholes for water. World Wildlife Fund itself became almost more of a development agency than one for wildlife conservation as it dealt with villages around the great reserve of Andohahela, which straddles the rainfall fault line from rainforest to spiny desert. Other NGOs also settled in for the long haul, particularly Japanese Southern Cross, which replants spiny forest cut for planks and charcoal, and ASOS itself, turning from famine relief to continuing health aid.

A spectacularly successful project began in the Upper Mandrare Basin. IFAD, the International Fund for Agricultural Development, made loans totaling $18 million from 1996 to 2008, with a further $8 million from other agencies. Long and bitter experience has shown that simply throwing money at a problem won't solve it. However, this scheme was planned with full farmer cooperation, and resident technicians devoted to their outreach task. To begin with, the World Bank rehabilitated the road that connected to the Fort Dauphin road, cutting the 300 km journey from fourteen hours down to six. Then the project tackled irrigation, water users' associations, improved rice planting, livestock vaccinations and veterinary care—even reforestation with Moringa, a local tree useful for everything from human and cattle food to wood to medicines. Local rice yields soared from 2 to 4.6 tonnes per hectare, and family incomes likewise. Sadly, after the coup of 2009, with the breakdown of law and order and the rise of cattle theft, many gains have been lost—but at least the 100,000 farmers now know how to prosper and that they can do so.[11]

At the end of the drought and famine of 1991–92, the Androy region was less isolated than ever before—practically a part of the modern world.

But, basically, the south is unsustainable. Spiraling population growth and shrinking resources mean more and more people live from food aid. More drastically, the future prognosis is not more of the same, but worse. Here is ten years later:

June 22, 2002, Berenty.

Moumini Oedraogo, tall, slim, round-faced World Food Programme rep from Burkina Faso, very round dark eyes, close-cropped round skull. He was a beneficiary of the Catholic Relief Services school feeding program as a child when he used to walk 8 km every day to school. Moumini later worked for Catholic Relief Services in Burkina, then for WFP: two years with refugee camps and displaced persons in Goma, general chaos in Kinshasa. Now here, based in Fort Dauphin.

In the famine ten years ago, David Fletcher et al. moved about 19,000 tonnes of food to the south in 1991, and then another 27,000 tonnes in 1992. It took two more years to taper off free food and move to food-for-work, with much soul-searching on WFP's part about continuing after the immediate emergency. Now the program is 5,000 tonnes per year.

The part Moumini is proudest of is the school feeding program. He had to convince WFP that if they were sending trucks out to give food for work, they could stop at schools on the way. There had been school feeding programs

11. B. Shapiro, A. Woldeyes et al., *Nourishing the Land, Nourishing the People: The Story of One Rural Development Project in the Deep South of Madagascar that Made a Difference* (2010).

before—started in 1975 for the whole country, then reduced to just Tulear province: the south. However, 'monetized food'—that is, food sold for money that was supposed to feed back into the program; in other words, money—just disappeared. Also five trucks given to Tulear were carrying anything else but food, while the food rotted in the warehouses. Donors bailed out in January 2001.

Now Moumini has three UN volunteers who take charge and a community monitor in each school so the teachers don't divert it, and reports each month on the number of boys, girls and cooks fed. Can't top up teachers' salary—that's a government responsibility.

Moumini's WFP group builds water wells. Traditional pits are narrow cones in the earth about 25 feet deep with a spiral path down the sides. These are built and maintained by the community. Cement versions can be done with WFP food-for-work. The Japanese instead installed hi-tech solar panels powering a pump which carries water to the town of Ambovombe, but much has rusted apart in the three or four years since installation. The basic problem is that there are two water tables: fresh water with salty water below. If you draw off too much fresh, the salty water rises and the system is then useless for people, though zebu may still drink it. (In Androy the zebu matter much more.) The locally built wells at least draw the fresh water very slowly. I saw a natural crack in the rock recently where they sent a teenaged girl down a kind of chimney 8 meters deep. There she could tip a coffee-can on its side to scoop up half a cupful of water at a time.

SAP based in Ambovombe is the *Système d'Alerte Précoce*, the new early warning system for drought. Observers in each commune fill in forms: one on agricultural output, one on prices in the market and whether there is distress selling of cooking pots, and a third one on social conditions in general. The observers get paid 15,000 FMG ($2.50) per form. The forms are collated and compiled into a preliminary report in April, with a final handsome published report in June. This year suggests bad needs in two communes but others are supposed to be OK.

The catch is that June is too late for most international agencies to respond to a famine due in September–December. Also SAP costs $150,000 per year. Expat salary and computers? Enough to buy an awful lot of food.

The most telling remark of all. Moumini comes from Burkina Faso, and has worked throughout the Sahel, in real desert. There they know how to conserve water: stone dams in fields, trees, storage wells. In southern Madagascar there is far more rain but people don't do anything about saving it, however much they suffer. In Madagascar he saw for the first time in his life people lying flat on tarmacked roads after rain, lapping at the puddles.

Moumini leaves, admitting that his wife and small baby are arriving *today*, in two hours, at Fort Dauphin airport, all the way from Burkina Faso, where she had the baby last July. He was there then, but has not seen her since! We said hasty goodbyes.

Eight years later than the last excerpt, eighteen years later than the 1992 famine, 2010 is another bad year. Another diary:

Jan. 18–19, 2010.

High-level Unicef–UNDP tour of Androy. I was shocked at sailing into Tandroy villages in a fleet of three colossal Land Cruisers with aerials for GPS and other communication sticking up to rival the Fantsiolotse spires of the spiny forest. A normal wooden Tandroy house is shorter than one of the Land Cruisers. This is the kind of mission I associate with the World Bank, not Unicef. The look that would make people in Afghanistan set off bombs.

SAP signaled an oncoming crisis in December 2008. In March 2009 Unicef started intervention. Active searching out of acutely malnourished children. Total of 8,713 kids treated in Androy and Anosy. 60 percent cured, 17 percent abandoned program, 2 percent died, about 19 percent did not gain adequate weight, 3 percent transferred. Lack of adequate weight gain likely due to sharing Plumpy'Nut ration among siblings as soon as the target child is out of immediate danger.[12]

We stop at Maoralopotsy, a tiny village and basic health center south of the main road. It has two new, very young nurses in white coats (med. techs). Cement building with blue doors, two rooms, in wooden village. About thirty women and children squat outside. The med. techs treat diarrhea, pneumonia, malaria. They also follow about 100 pregnancies out of 500 in their area. Only three or four women per month deliver here, the rest with village midwives. Many people live 7–12 km away and in labor would have to travel for two hours over the potholed tracks by ox cart to the CSB. If there are complications they go on by ox cart more hours to the hospital at Ambovombe. Open boxes of birth control records sat on a dusty wooden shelf, so, while the others pored over the CSB nutrition register, Peter Metcalf, (the country's UNDP rep.) and I picked out a few of the yellow cards. Peter remarked, 'So much for medical confidentiality.' Of the 32 records this month, only one woman had as many as eight children. Most of the records were for young teenagers with no or one or two

---

12. Plumpy'Nut is a kind of peanut butter or Nutella fortified with milk powder, vitamins and minerals. It comes in small plastic packets which keep for two years without refrigeration. A commercial product made by the French Nutriset company. It tastes good!

kids, the balance for women in their late thirties at the end of childbearing. This is already a huge change.

❦

The fundamental coping mechanism is the family. Barrenness is the ultimate curse—and not uncommon, given the prevalence of gonorrhea. With a large family, some will support the rest and the aged parents… Besides, leaving descendants is the greatest pride for any Malagasy. Westerners may tut-tut, but it is a rational (and joyful) strategy. As my friend Lahivano said when being harangued about birth control, 'You don't understand. We *like* children!' The family unites to celebrate funerals. Funerals involve feasting and dancing and sex between those cousins distant enough for it not to count as incest. Funeral gatherings and the exchange of brides link ever more extensions of the family. Family is the only sure recourse in the recurrent emergencies.[13]

The other fallback is the herd. At great funerals, zebu brought by each son-in-law stampede down the village main street while men urge them on with sticks and blank gunfire—round and round through the village to make the herds seem inexhaustible. Of course each owner knows exactly how many he owns, which means that the boasting songs about the number of zebu and the number of grandchildren must be finely judged, because other people know too. In the old days, when there was more forest for pasture and fewer people, a man's whole herd would be slaughtered at his funeral. Missionaries tut-tutted about this apparent waste of resources. Nowadays most people are nominally Christian, which means only a few of the zebu are killed: enough to feed

13. Incest avoidance involves complex calculation by paternal generations. Since a man may have several wives at once or in series, his oldest son may be 40 when his youngest is born, but the sons count as the same generation and their offspring as first cousins. For the role of daughters-in-law, see K. Middleton, 'The rights and wrongs of loin-washing' (2000).

the scores or hundreds of invitees and to send off the dead ancestor with the nucleus of a herd for his afterlife.[14]

Depending on livestock, rather than crops, makes sense in a drought-prone area—sheep and goats and zebu live much longer than the shriveled stalks of maize. However, when even livestock die (or, in the modern day, when the market price falls to nothing because everyone is trying to sell), then the destitute still must turn to the extended family.

A final buffer used to be the forest. The commonest famine food is tamarind pulp slaked with chalk to cut its acidity. People tell me that tamarinds will keep you alive for a month or two, but with recurrent bellyache. There are also edible forest roots and tubers. The forest, of course, is adapted to recurrent droughts and still is green when the fields lie burnt and bare. However, there is less and less forest—and the deforestation rate of spiny forest is now some 3.2 percent per year, the highest in Madagascar. The cleared land is turned into fields or charcoal at roughly the same rate as the southern population grows.

Later we will return to the situation after the coup in 2009. Cattle rustling out of the region has skyrocketed: the pride and the family investment in the herds have almost gone.

Bottom line: change is coming, and must come. There is no way the south can return to a pastoralist lifestyle, to mining the spiny forest, to population control through disease, famine, clan war. And yet there is no prospect that agriculture in this dry land will do more than feed local people—in good years. People always, or almost always, find new ways to cope, but how long that will take is shrouded in the heat haze of the future.

---

14. A. Jolly, *Lords and Lemurs: Mad Scientists, Kings with Spears, and the Survival of Diversity in Madagascar* (2004).

TWENTY

Lemurs coping

Ring-tailed lemurs are female-dominant. When I say this in lectures, some guy in the audience is likely to wise-crack 'just like us!' while several women punch their fists into the air.

Not like us. No adult male can challenge an adult female, even if he is the swaggering alpha in the male hierarchy. Male ring-tails have spurs on their forearms which gouge scent into saplings or perfume their bushy tails. Some of them walk like John Wayne. If a female thinks a male is too close or if she wants the tamarind pod he is currently holding, she cuffs toward him or actually biffs him on the nose. He has no choice but to retreat giving a 'spat call,' a high-pitched series of squeaks that roughly translates as 'I am so sorry!' or at highest intensities 'Please pardon my existence!' Males cannot even force females to mate, though the female may well choose a winner in the males' kung-fu jump-fights of the mating season.

Female dominance is widespread, though not universal among the 50–100 species of lemurs, depending on how you define a species. Many more than a hundred different populations exist around Madagascar. In any one forest you may find from five kinds up to a dozen: one or two each of mouse lemurs, dwarf lemurs, leaf-eating lepilemurs, up in body size through the classic 'true' or brown lemur group, ring-tails, and the leaping sifaka and indri. Most lemur species are female-dominant, but that is not their only peculiarity.

Lemurs, and for that matter all the Malagasy mammals, have the wrong number of babies.

If you plot a graph of body size versus litter size, mammals fall into two distinct groups. Some, called altricial, pop out lots of blind, deaf, fairly naked, and to us unappealing, young that need a sheltered nest: think of rats or mice. Or, to be kinder, tiny kittens. The other lot, the precocial ones, have one well-developed young or at most two at a time. Think horses, or for that matter wildebeest calves, which must run away from lions on the first day of life. Primates are precocial, born furry with open eyes and ears and with a grip strong enough to hang on to mum as she ricochets through the trees. As for humans, our large brains mean we are born in an almost fetal state compared to our nearest relatives, so we have partially reverted from the normal primate condition to needing a nest of sling or carrycot.[1]

Now re-plot the graph of body size to separate the altricial and precocial strategies. Lemurs don't fit anywhere. Tiny mouse lemurs almost work for the lowest-size category: they have litters of twins in a nest, and can if lucky produce four young a year. But a house mouse the same size can have a litter of twelve, and even wild deer mice have litters of five. When you get to the larger-sized lemurs, they grow even more slowly for body size compared to other precocial primates, taking three or four or even six years to mature, and then usually raise just one infant a year.

Slow maturation and slow reproduction are not confined to lemurs. John Eisenberg pointed out back in 1981 that almost all Malagasy mammals are suspiciously slow reproducers

1. R.D. Martin, *How We Do It: The Evolution and Future of Human Reproduction* (2013).

(except for a few, like the spiny tenrec, which go to the opposite extreme, with a maximum recorded litter of thirty-two). Later work only confirms Eisenberg's suggestion. Something about the climate of Madagascar leads most mammals into what seems at first like overcaution.[2]

Two primatologists have converged on the same explanation. Alison Richard turned to her long-term records of the demography of white sifaka at Bezà Mahafaly to analyze 'life in the slow lane.' The sifaka population may lose the majority of a year's infants during drought years. A female takes six years to mature, reproduces every one or two years, and may live to 15 or even 22. They are 'bet-hedging.' If the mother is having a good year she may raise a strong, healthy infant; if not, and the infant dies, at least she does not squander so much energy that she risks her own life and future reproduction.[3]

Patricia Wright of Ranomafana called her version of the idea the 'energy frugality hypothesis.' Same fundamental reason: the hypervariable climate of Madagascar, not only from season to season but from year to year. Pat, however, extended her speculations to include much other behavior, including low metabolism, female dominance and female territoriality. In Ranomafana predation by fearsome fosa and infanticide by immigrant male sifaka mask some of the year-to-year effects, but long-term study shows that, even in wet rainforest, climate swings can take out whole cohorts of young. There the problem is more usually cyclones than drought but the effect is similar. Females and even males

2. J.F. Eisenberg, *The Mammalian Radiations* (1981); S.M. Goodman and J.P. Benstead, eds, *The Natural History of Madagascar* (2003); R.E. Dewar and A.F. Richard 'Evolution in the hypervariable environment of Madagascar' (2007).

3. A.F. Richard, R.E. Dewar et al. 'Life in the slow lane? Demography and life histories of male and female sifaka (*Propithecus verreauxi verreauxi*)' (2002).

conserve their energy against the shortages they have evolved to expect. A whole fauna of pessimists.[4]

Come back to female dominance. Does this mean that female lemurs have higher energy needs than males? I put forward this idea, which has been thoroughly shot down. The final nail in the coffin of exceptional female need was provided by my friend and closest colleague, Prof. Hantanirina Rasamimanana. With a raft of her students from the École Normale Supérieure, the university-level teachers' college, her crew counted every lemur bite of every plant species at all seasons of the year, and each leap and promenade, to see how much energy the lemurs expended. Males and females do turn out to have slightly different diets. Females eat more fruit; males, who often have to wait their turn in the choicest trees, have recourse to coarser leaves... but the males eat for longer and gobble faster. Much the same energy intake and output.[5]

I have told in another book how Hanta and I first met. Back in 1983 she was an intern in my very first Earthwatch group when she was already a qualified nutritionist working at Tsimbazaza Zoo. A huge storm loomed over the river, swinging lightning bolts like a cane. I ran through the forest

4. P.C. Wright, 'Lemur traits and Madagascar ecology: coping with an island environment' (1999); S.T. Pochron, W.T. Tucker et al., 'Demography, life history and social structure of *Propithecus diadema edwardsi* from 1986–2000 in Ranomafana National Park, Madagascar' (2004).

5. A. Jolly, 'The puzzle of female feeding priority' (1984); M.L. Sauther, 'Resource competition in wild populations of ringtailed lemurs (*Lemur catta*): implications for female dominance' (1993); M.L. Sauther, R.W. Sussman et al., 'The socioecology of the ringtailed lemur: thirty-five years of research' (1999); H.R. Rasamimanana, V.N. Andrianome et al., 'Male and female ringtailed lemurs' energetic strategy does not explain female dominance' (2006); B. Simmen, F. Bayart et al., 'Total energy expenditure and body composition of two free-living sympatric lemurs' (2010). However, sifaka female body weights do correlate more closely with ecological variables than male weights do, suggesting that females are the 'ecological sex.' A.D. Gordon, S.E. Johnson et al., 'Females are the ecological sex: sex-specific body mass ecogeography in wild sifaka populations (*Propithecus* spp.)' (2013); B. Simmen, F. Bayart et al., 'Total energy expenditure and body composition of two free-living sympatric lemurs' (2010). S.T. Pochron, J. Fizgerald et al., 'Patterns of female dominance in *Propithecus diadema edwardsi* of Ranomafana National Park, Madagascar' (2003).

calling out the volunteers, most of whom were already on the path. I couldn't find Hanta, deep in the woods. At last she answered my call. The two of us raced back up the main trail as the storm struck and branches crashed all around us. 'Didn't you see the sky?' I panted. 'Of course, but you said to stay with our troops, even if it rained.' I decided this was a very special lemur-watcher. It was not till years later that she teased that I had never twigged that she, like other Malagasy scientists, were instructed by the paranoid, socialist government to spy on international visitors and report what skullduggery we were actually up to in the woods! I also never expected that we would become close friends, that her sons would spend terms at our home in Lewes learning English, or that her careful work would pop my pet hypothesis on lemur female dominance.[6]

However, I now see a different link between female dominance and the hypervariable climate—through territoriality. Females defend their own or their group's territories, not males. A feeding territory is a long-term buffer against the hardest years and seasons. Most importantly, females who manage to hold on to their territory pass it on to daughters and granddaughters—the ring-tail troops in most Berenty areas are the successful descendants of owners decades before, though many unsuccessful matrilines have been evicted. Indeed, females may continue to fight for their property in a season when raiders are actually getting more food from it—long-term ownership is worth short-term effort. Long-term means not just year to year, but generation to generation. Males, as in most primate species, change troops as they mature and at intervals thereafter, so they have much less stake in owning property.[7]

---

6. A. Jolly, *Lords and Lemurs: Mad Scientists, Kings with Spears, and the Survival of Diversity in Madagascar* (2004).

7. A. Jolly and R.E. Pride, 'Troop histories and range inertia of *Lemur catta* at

This is only a hypothesis. Not all lemur species or even populations are overtly territorial. An in-depth review of all the literature on lemurs and other territorial mammals would prove or disprove the idea, particularly if linked to social system and population density. (I am not about to do this, now.) If there is masses of space, and masses of random predation by fosas or others, there may be little occasion for individuals to meet unless intent on mating. There you don't see much competition of any sort. Wherever they are squeezed together, though, lemur females can be ferocious.

Almost all research on ring-tails and the southern sifaka has been done in two small reserves. These are dense, shady gallery forests beside rivers, with tongues of spiny forest as one leaves the riverside. We'll talk of the human side of Berenty and Bezà Mahafaly in the next chapter, but here is a little of what can happen to ring-tailed lemur behavior in a bad year.

How do ring-tailed lemurs respond to droughts? It depends...

At Bezà Mahafaly, 50 percent of forty-eight original females died in the drought of 1991–92, and 80 percent of infants in 1992. The population did recover afterwards. A few remnant troops joined up together. Several females gave birth to twins, and they were back to their original numbers by 2001. Lisa Gould, who led the Bezà study, concludes that ring-tail life history includes crashing and rebounding and relatively (for a lemur) rapid reproduction. This was clearly true in the somewhat harsher conditions of Bezà, but not so in the milder conditions of Berenty.[8]

Berenty: a 33 year perspective' (1999); A. Jolly, H. Rasamimanana et al., 'Territory as bet-hedging: *Lemur catta* in a rich forest and an erratic climate' (2006); R.E. Pride, D. Felantsoa et al., 'Resource defence in *Lemur catta*: the importance of group size' (2006).

8. L. Gould, R.W. Sussman et al., 'Natural disasters and primate populations: the effects of a two-year-drought on a naturally occurring population of ringtailed lemurs *(Lemur catta)* in southwestern Madagascar' (1999); L. Gould, R.W. Sussman et al., 'Demographic and life-history patterns in a population of ring-tailed lemurs (Lemur catta) at Bezà Mahafaly Reserve, Madagascar: a 15-year perspective' (2003).

At Berenty troops split. Infants died in the mayhem as well as from simple hunger. The two troops we knew best wound up in a lemur war between competing matrilines. However, the fifty females in the natural part of the forest, away from the tourists, stayed about the same number: a stable adult population, even while first-year mortality went up from around 25 percent to about 50 percent. Meanwhile, at Bezà in the same period, the larger and longer-lived white sifaka reacted like the Berenty ring-tails, losing much of the infant cohort in harsh years, although most females survived.[9]

To put it bluntly, life histories of Berenty ring-tailed lemurs, Bezà white sifaka and the Tandroy of historic times are disturbingly similar. Catastrophic drought will happen in your lifetime. Even though mothers of each species love and cherish their babies, the first priority must be to survive yourself. Only then will you be able to look forward to eventual grandchildren.

When a ring-tailed troop is under pressure dominant matrilines may target subordinates to evict them from the feeding territory. In the birth season, when females use up the last of their stored energy and feed on what fruit is available before new leaves and flowers unfold in anticipation of the rains, social tensions may erupt into aggression. Not directly lethal aggression, though if the targeted animals hung around it clearly would be. Instead, the lowest matriline is chased away to fend for itself.[10]

This can, in fact, be indirectly lethal if the evicted females cannot find unclaimed land, fight to partition the original home range, or oust someone else's troop to take over a territory. Just once have we have seen the underdogs win—and

---

9. Richard, Dewar et al., 'Life in the slow lane?'; Jolly, Rasamimanana et al., 'Territory as bet-hedging.'

10. L.G. Vick and M.E. Pereira, 'Episodic targeted aggression and the histories of Lemur social groups' (1989).

that was in the drought year of 1992. You can find statistics in the papers, but most scientists have to be too careful of their careers to write what it feels like to watch...

Monday, September 21, 1992.

Lemur War. A1 and A2 are new troops. The parent group, the A Team, split no more than a month ago. A1 has five adult females and two 2-year-old daughters, outnumbering and outranking the four females and only 2-year-old in A2.

The A2 females have always been the subordinates: that's why we assigned them '2' and not '1'. Last birth season the subordinate females nearly lost babies three different times, just from being pushed around so they couldn't get back to retrieve fallen infants. Last year, though, they acted as though they would do anything to be allowed to stay in the troop. They meowed pitifully to the dominants and hung back waiting their turn for everything important, like food trees and water sources. This year they are physically chased out of the troop.[11]

To survive now, the A2s are going to have to mark out a territory—they can't live at Berenty without some rights to home ground. And they will have to do it in contest with females who have dominated them for many years. I don't have much hope for them. So far no males have committed themselves to the new troop for more than a few hours' visit. Males don't seem to be much help in fighting—they just perfume their tails with their wrist glands, in case there are other males to impress. It isn't a good sign that the males are also hedging their bets. Like us they are waiting to see if the A2 females are tough enough to last.

Princeton seniors Laura Hood and Sharon Katz get up at 5 a.m. in the first light of dawn. They trek down from their

11. Gwendolyn Wood, pers. comm.

little cottage through a forest ringing with birdsong. Students aren't housed like tourists: their floor is riddled with termites and their kitchen is full of giant flying cockroaches. However, they have the privilege of walking in the magic forest. Couas squawk at each other, flicking their feathered crests, the sunlight glinting from their turquoise and rose madder eyepatches. White sifaka, larger cousins of the ring-tailed lemurs, bound through the trees like ballet dancers—or on the ground like children in a sack race. Even the chameleons and boas and emerald-and-scarlet day geckos are unique—because this is Madagascar's alternate world of evolution.

Laura and Sharon pick up A1 and A2 in the forest. Both troops sleep there, high in the massive branches of tamarind trees. The tamarinds are just starting to turn gold. Our autumn color comes two weeks before spring: the tamarind leaves turn, and fall, and then the new delicate pink leaves emerge. Incongruously delicate on trees 60 feet high and 50 or 60 feet broad. Tamarinds grow like the massive white oaks of northern woodlands. Of course, September is the southern spring.

The two rival troops come through the forest in parallel, leaping from swaying liana to tamarind bough. They pause to nibble a flower or a new leaf in passing, but they are heading out to the part of their range that they share with humans—and in particular one dense tamarind tree that grows all by itself in the cleared field beyond the tourist bungalows. A2 clatters across a thatched roof. A party of Swiss emerge with telephotos, cooing, '*Wie shöne!* Oh look at the baby!'

The Swiss don't notice Sly. Sly is wounded. In a fight two days ago she nearly lost her right ear. Her scalp is gashed open—not to the bone, because her ear still twitches, but with a red two inches where the skin is flayed and muscle shows like an anatomical diagram. Her silhouette makes A2 easy to recognize at a distance: the one ear vertical, the other out flat.

Laura, who knows all the lemur faces of A2, can teach them to me. She turns schoolteacher. 'Which one is Shadow?'

'Shadow is probably only 3 years old—she's the only one who still has a baby. Beautiful unblemished coat. That's the infant we watched being born in the forest.'

'How do you tell Blotch?'

'Darker head cap and narrower forehead star than Shadow.'

'Who is left?'

'PJ and Sly. PJ has a dark muzzle with a brown-colored 'bag' under one eye. Do you know, I finally asked Dr. Koyama what 'PJ' stands for? PJ was young and elegant eleven years ago when he first studied here. PJ is 'Parisienne' with a heavy Japanese accent. Sly almost lost her baby last year just from being picked on. Her infant this year is overdue.'

'Hey you guys—Hey guess what happened!' Sharon called from out in the field. 'Sly just had a baby when I wasn't looking! I only lost track of her in the leaves for fifteen minutes, and here she is cleaning it off!'

Later.

At 5 in the afternoon there was an almighty racket from ring-tails, screaming 'war cries' that I only hear when groups are fighting. Helen Crowley, the Australian Reserve manager, came to find me. 'Sly has lost her baby.'

It lay a long way out of any area that A2 had been allowed to trespass on. Baby lemurs are tough. It was only ten hours old, but it was wriggling and beeping and clutching, the umbilical cord still attached. It would normally hang upside down on its mother's belly as she leaped through the trees or fought other females. It lay on its back on the ground, clutching at all it could find—its own tail.

A2 and A1 had fought on what is crystallizing as their frontier—but when the baby fell off, Fish, second dominant of A1, let it crawl onto her, along with her own infant. When the two troops recoiled to their own areas, Fish ran with her own side. The baby hung on for twenty minutes, long enough to be carried hundreds of meters away from the region A2 were trying to stake out as their own. Fish's semi-maternal response condemned Sly's baby. Sly couldn't even hear it beep from where it lay.[12]

'I don't care,' I decided, 'I am not going to kill that infant. We've given back fallen infants before, outside of the study troops. Let's at least give Sly the chance to have it back.' I scooped it up onto a clean check-sheet on the chance that would smell less human than sweaty hands.

We carried it quickly to where A2 was recuperating from the fight. Sharon was still faithfully taking notes. We laid the infant on the ground, where it beeped again and again. Blotch and Sly descended to the call—any female is likely to come to any crying infant. Normally a mother heads straight to her offspring, and hovers above it so it can grab her fur, even if she does not scoop it up with a hand. Instead Sly just stood well back, and stuck out her nose to sniff the crying baby as though she had never smelled it before. Blotch did the same. You couldn't have told, to look at them, which was the real mother, or if either was.

They leaped away leaving the infant there. It cried again. Sly sat 2 meters off, looking back and listening. Her troop

---

12. Ring-tails do actually kill infants, though very rarely. In the sparse record, females are more likely than males to slash another female's infant with their canines. A. Jolly, S. Caless et al., 'Infant killing, wounding, and predation in *Eulemur* and *Lemur*' (2000).

was moving into the forest, starting the last trek of the day toward their sleeping site. Sly meowed her contact call at the departing troop, and followed them into the shadows.

An hour later Michael Pereira from Duke Primate Center turned up at my bungalow. 'I've been telling the students to expect less maternal behavior,' he said. 'These females are fighting for their lives, so it may be worth their giving up this season's reproduction, to save energy for fighting.'

The abandoned infant died in the night.

Tuesday, September 22, 1992.

Something has happened to Sly's behavior. A1 and A2 have had yet another fight by the tamarind tree in the field. It started out like all the others: Shadow and Blotch in front, PJ and Sly behind, with the two most aggressive of the A1 females in the front rank for their side. Every other time, though, Sly and PJ have been meters away or come tentatively forward, only to slink off in mid-confrontation to a farther tree, leaving Shadow, Blotch, and the 2-year-old Possum to retreat as best possible without turning their backs and being jumped from behind.

Today, Sly is also almost in the front rank. The aged PJ is only a little behind. Laura points it out, surprised since she has seen so many confrontations with the old lineup. Does losing her infant mean that Sly has no more reason to be careful? But Shadow is always the champion, child and all.

Sunday, September 27.

This is Sunday. Laura and Sharon and Michael have gone away for a long delayed weekend at the beach, as well as to shop for more groceries to feed their kitchen cockroaches. I like being here on Sunday—no tourists allowed. Maids abandon their white smocks and guardians put up their spears to appear in clean shirts and dresses and earrings, and walk up to church. There they will sing hymns in multi-part missionary harmony.

I promised to keep an eye on A1 and A2 just to make sure there are no more developments—that is, that everyone survives the weekend. A1 comes out of the forest to sun on a thatched roof. Where is A2? Over there in a tamarind, white bellies facing the sun, arms out, in the attitude the Malagasy say is praying. They are so photogenic I grab my camera and am snapping my several thousandth picture of ring-tailed lemurs before my mind registers why this is a different picture.

Seven lemurs sit in contact, or almost in contact, on one branch: all four females and the 2-year-old and the two 1-year-old juveniles. They don't take up as much as a meter and a half of branch. I could understand it if they were huddling in cold, but the distrust between lemur females, even within a troop, is usually too strong for them all to sit in quite such a tight band for no good reason. A2 looks like a picture of togetherness.

I hear mewing at my heels. I look round. A1 is ranked behind me, headed for the rich pods of that particular tamarind tree.

The A2 females don't wait. They leap out of the tree—but for the first time they are the attackers. Shadow is in front but all five are in the fray, grappling and lunging, the dust flying up as lemurs spin over each other or break to face new opponents. Females rarely bite each other: they parry and feint, grapple and break too quickly to reach each other often with a slash like the one on Sly's skull. Almost always two or three females of even the largest troops do the actual lunging for them all. But in this extraordinary moment all five A2s attack at once. They check A1, which had dominated them for so many years.

The combatants break and glare at each other. Only two of the A1s are actually involved—the rest try a flanking movement behind the bungalow toward the tamarind. Shadow sees them, flings herself round the corner of the house, and drives the others back. Then the A2 females occupy the thatched roof while the A1s clamber up a little flamboyant tree and feint as though they will attack. The row of A2 faces glaring from the roof like five live gargoyles with canines gives the others pause—even Sly is there with her half-mast ear. Shadow launches herself from the roof. (All this time her infant—the one surviving A2 infant—clings upside down to her belly fur.) A1 now has five infants—they do have reason to defend this year's babes. Shadow lands almost on top of A1's front-rank defender. The little tree dips and sways, and it is too much for A1. They back off—literally backing—this time they do not dare turn away. And then at a safer distance they promenade away down the path, at a stately pace that fools no one. Shadow and Blotch and Sly and PJ and Possum, with due caution, move into the next trees—a eucalyptus grove which they had never reached since the troop split.

I left them there gobbling eucalyptus buds. The grove was loud with bees, the white-stamened flowers dusted lemur noses with pollen-like powder puffs, and the conical ivory bud caps pattered on the ground. All around A2, the air smelled of honey.

October 30, 1992.

It has almost rained! Big separate drops land upon the dust, sending up little puffs of soil, and disappearing instantly into lower layers that are dry as talcum powder. People wander round with big grins. Rekanoky, the headman, and his young wives and little daughters all saunter along grinning. Earthwatcher volunteers, come to help me track the troops, stand outdoors looking up, their fancy wildlife T-shirts a trifle damp. The grey sky looms behind the yellow leaves remaining on the tamarind trees.

Our colored tree-flagging tape tracks A2's progress through the eucalyptus grove, day by day claiming a few more trees. Now it is A1 who are in trouble, retreating as A2 appear. But the lemurs don't like rain, even after two years' drought!

A2 take shelter on the sign of the reserve gate. Its peaked roof is meant to protect the paint from the weather: pictures of ring-tails and sifaka and flying foxes, rather crudely drawn… Neatly curled under the roof sit Shadow and Blotch and PJ and Sly, with Possum the 2-year-old and the two juveniles. Seven animals in a row, all tails over shoulders like an endless feather boa, their fur in little wet tufts, and their noses buried in each others shoulders. Seven furry little balls all in a close-cuddled line. They still don't have full-time males, but A2 are going to survive.

TWENTY-ONE

Scientists, people, lemurs: Berenty, Bezà Mahafaly and Tsimanampetsotsa

What of conservation of the spiny desert and its fauna? Three different reserves give a counterpoint to Ranomafana, not just in landscape, but in outlook. Berenty remains an argument for feudalism. Yes, I did say feudalism. Bezà Mahafaly shows what can be achieved by outsiders' endless patience to respect local people, however different their mindset from one's own. And Tsimanampetsotsa gives the near-impossibility of working at large scale in a lawless society.

Fifty years ago, in 1963, just after finishing my Ph.D., I stumbled upon Berenty: a patch of gloriously shady gallery forest beside the Mandrare river. Berenty's animals had been protected since 1936, a year before I was born. Everywhere else in Madagascar I was lucky to see tail-tips of fleeing lemurs, rightly fearful of humans. Here they came to me instead. The white sifaka leaped in my direction chorusing their alarm call *Shifakh, Shifakh*: a sound to American ears halfway between swearing and a snore. Sassy ring-tailed lemurs encircled me with ear-splitting soprano yapping, their own mobbing vocalization.

The nearest town to Berenty is Amboasary, on the main and often impassable road from Tana to Fort Dauphin.

❦

Jan. 7, 1963, letter home to my parents in Ithaca

Amboasary lies in a dusty, rainless valley. To the east rise the coastal mountains, capped with white cloud, cut off as though with a knife between mountain and valley. The mountain is rainforest. The valley is the weird Didierea forest, with thousands of horned and fleshy plants like cacti, all endemic to Madagascar. Attenborough's television pictures show you the 30-foot vertical spikes of *Alluaudia*—I think he does not show the helical rows of lobed leaves, with a thorn between each leaf. How the Propithecus can leap those plants without wearing armored mittens is beyond me.

But the *Alluaudia* is beaten and retreating. When you come to the slope down to Amboasary, miles upon miles of blue-green sisal spread like a lake in the plain. It grows in a geometrical fan, in geometrical rows. As I was trying to think of descriptions to fit those thousands of spiked leaves all pointing up and out in absolute relation to each other, Preston [Yale student assistant] said: 'I'd hate to make a forced landing here in a canvas-bottomed airplane...'

Berenty Reserve lies in the midst of Madagascar's first great sisal plantation. The owners are the de Heaulme family, French Reunionais aristocrats. The rich, shady gallery forest of giant tamarinds is still there because the paterfamilias, Henry de Heaulme, declared that it was too beautiful ever to cut down. Remaining spiny forest on the estate is full of life as well, protected from zebu, goats and woodcutters. In all, 1,000 hectares of the 6,000-hectare estate remains in forest, with some connections to the much scrabblier forest outside.

The forest reserves now earn their keep. The canny owners did not listen to pessimists who chorused 'Foreigners go to Africa for elephants and lions. Why would they ever bother to come see *lemurs?*' Berenty became the first ecotourist destination of Madagascar back in the 1980s. Now tourism is the first or second contribution to Madagascar's GDP, at least in years with no political unrest.

Nowadays, followed by scientists and photographed by generations of tourists and TV, the Berenty animals are far too tame to worry about people. They mostly go about their business, dancing for the cameras or holding aloft their plumes of ringed tails.

The people of the Mandrare valley are clans of the Tandroy tribe, the 'People of the Thorns.' It was not for decades that I felt I must look beyond the forest to the sisal workforce, the original owners. Then I crystallized a view of multilayered societies in intersecting orbits, each maintaining its own traditions. Six lemur species. A hundred kinds of birds. The French owners, the multilingual tourists, the ever-growing Science Tribe. And the Tandroy.

Berenty's Tandroy sisal workers are buffered against famine, not only by their meager salaries, but because the de Heaulmes were careful to leave space for local farm patches so people could supplement their own food supply. On some other plantations the owners cut from end to end of their colonial government 'concessions,' which leaves almost no recourse if workers are laid off. Tandroy traditions are strong: if you are lucky, friends may invite you to a healing and exorcism ceremony, with its ululation, days of dancing, and a final sacrifice of goats. (Cattle are saved to sacrifice at grand funerals.)

Probably the greatest contribution of Berenty to conservation has been the Science Tribe: Americans, Japanese, Germans, English and, especially, the continuing flow of eager students from the École Normale Supérieure trained by Hanta Rasamimanana. Her leadership and example have launched a generation of young scientists and conservationists—not for Berenty itself, but for all Madagascar.

Berenty Estate has also conserved a significant forest, the largest bit of gallery forest on the whole Mandrare river. But it is hardly a model, unless you have owners stubborn enough to resist every political pressure of the last fifty years. As Jean de Heaulme remarks, 'Short-term conservation never works. What you need is to build a country where you would like your grandchildren to live.'[1]

This is also in part the message of a reserve founded on exactly opposite principles, Bezà Mahafaly. In 1974 an unlikely triumvirate set out to find a community which would protect a forest and open it to researchers from the University of Antananarivo. The three were Alison Richard of Yale; Bob Sussman of Washington University, St. Louis; and Guy Ramanantsoa, chair of the Water and Forestry Department in ESSA, the Agronomy school of the University of Antananarivo. They were backed by a visionary administrator: Gilbert Ravelojaona, president of ESSA. Not just visionary: courageous. Hard-line socialism in 1972–75 cancelled research visas for foreign scientists and brought general paranoia about foreign designs on anything, let alone conservation. When Ravelojaona said he would sign papers for a university-led reserve involving foreign professors, he was courageous indeed.

Guy prospected throughout Madagascar. He eventually traveled by ox cart to meet the mayor of Ankazombalala (former Beavoha) Commune, who offered the foreigners a cautious welcome. To a cluster of villages in the deep south-west, 'foreigners' includes anyone from the capital. The account that follows is almost wholly from the 37-year summary by

1. A. Jolly, *Lords and Lemurs: Mad Scientists, Kings with Spears, and the Survival of Diversity in Madagascar* (2004); A. Jolly, 'Berenty Reserve, Madagascar: A long time in a small space' (2012).

Alison Richard and Joelisoa Ratsirarson of the extraordinary partnership that Bezà has now become.[2]

In the first phase, of about ten years, the 'partners' had very divergent ideas of what their reciprocal agreement meant. The universities' goal was conservation and research; the communities' goal was economic return. Deeply rural communities might well be mystified by the researchers—but the researchers learned that a community is an idealized reification of lots of different individuals. The 'foreigners' dealt with the mayor and other prominent people, but there were many others, including one whole village that never accepted the reserve and continued to make clearings and pasture cattle. And oops: the chosen name was the mayor's village, actually 10 kilometers away from the reserve, not the nearest village nor the overall commune.

The University was not a development agency. Funding for local improvement was scarce, though there was help from WWF–US. The villagers complained that they'd given up 600 hectares of forest for little visible return.

Then came 1985, with its 180-degree change in national conservation policy. Bezà was adopted as a Special Reserve under the Ministry of Water and Forests. Money would flow from USAID. The minister of Water and Forests himself trekked all the way to Bezà Mahafaly for the inauguration.[3]

November 1, 1985. Tulear to Bezà Mahafaly Research Station.

The jamboree to open the new reserve! Morning start in 25-seater Mercedes bus. Russ Mittermeier's handsome face now has handsome deep laughter lines—he'll probably look thoroughly leonine as he ages. Eleanor Sterling: California

2. R.W. Sussman, A.F. Richard et al., 'Bezà Mahfaly Special Reserve: long-term research on lemurs in southwestern Madagascar' (2012); A.F. Richard and J. Ratsirarson, 'Partnership in practice: making conservation work at Bezà Mahafaly, southwest Madagascar' (2013).

3. J. Randrianansolo, the 'Napoleon' of Chapter 7, 'Napoleon versus the Zoos.'

Yalie in her early twenties, working for WWF, inspired by Alison Richard to try for a Ph.D. on the mysterious aye-aye. Professor (Le President) Ravelojaona of the Agronomy School—tiny, limping from birth, great man of conservation. Witness the affection and respect of the colleagues which surround him, including his former student the minister. Forester Joseph Andriamampianina has matured from his earnestness of ten years ago (or indeed from the young forester who gave me my 1962 research permits) into smiling serenity—his face is younger and smoother as his close-cropped hair grows whiter. Pothin Rakotomanga, collaborator at the ESSA Water and Forestry Department has masterminded all the arrangements at the Reserve itself. I somehow imagined him as young and eager because he seems to be Alison's right-hand Man Friday—he is, in fact, gray-haired and perpetually happily smiling. Roland Albignac gives discourses on everything as we start, mostly on his monograph on carnivores, for Russ who writes it all down in his notebook. I see a lot of my own style in Roland. Renée Wynn, extraordinarily chic in straw toupee and almost non-existent T-shirt, gold necklaces, khaki shorts and bamboo cane. Why do the rich even tan more richly than the rest of us? But no riches would compensate for having to live on daily courage with multiple sclerosis. World Bank. USAID. Voluble reporter from *Madagascar Matin*—at least voluble at midnight, admitting that he is of peasant stock and knows well the peasants' love of fire which burns up grass and trees and government decrees.

Stopped at charcoal burners on the Tulear Road. Tulear town is surrounded by wasteland where there used to be forest: the town has no fuel besides charcoal. Charcoal burners' houses look as though erected yesterday for an overnight stop. They are straw-and-stick rectangles, mudded on the outside and inside, with a thatched roof that overhangs to shade the walls from direct sun. When the charcoalers first came twenty years ago, there were forests round the village—now it is bare savannah with the odd relict baobab. The woods where they cut are 6 km away, the kiln 4 km. They take ten days to dry the wood, and sell it for 1,250 FMG a sack (about 2 bushels), which is resold in town for 1,900 FMG a sack. One kiln produced the pile we saw—say ten to twenty sacks, or 12.50 francs ($19.34) for ten days' work, well over minimum wage—except this is for the whole village. They stay here instead of moving nearer forest because it's fertile land, so their crops grow.

The most revealing remark was the minister's—that the charcoal was excellent, thick in diameter. Trees that thick in this dry climate are primary forest, many decades old at least.[4]

The bus driver took his chance to change a tire, so Russ took his chance to shoot four rolls of film of the village. Russ persisted in wearing a kind of space-age mercenary costume—metallic silver singlet and shorts, with a South American machete hung at his side. (Later he embellished the costume with two local spears.) He was hot, the tripod over his shoulder burning hot. The minister asked, 'Who is that man who looks like a film director?'

Rolling on, through savannah nude except for fire-resistant *Sclerocarya caffra* (the thick skinned Kaffir tree) and the fan-leaves of *Medemia*, that palm which traces invisible watercourses. Over the valley of the Onilahy river: dry limestone bluffs round the wide plain. In the pluvial a few thousand years ago, was the Onilahy, wider and deeper than the Hudson, majestic between its lush borders?

---

4. Later, Conservation International's satellite maps showed the area around Tulear as the most extensive recent forest devastation in Madagascar.

After Betioky town came a dry watercourse—where predictably the bus with its twenty-five passengers stuck in sand.

It was pulled out with half the dignitaries pushing and joking and half, including me, on tow rope. (Not the minister, who loftily surveyed the scene.) It was galling that two ox carts trundled slowly past us down and up the sandy banks and across the dry stream.

At last to the reserve—the first *Alluaudia* and proper spiny desert we've seen! Does it not grow higher up in the dry bush country, or was it all cut for house timbers? Campsite—low, long, thatched shelter with picnic tables for sixty, and space for tents in the dust. A fridge with cold beer under the thatched campsite verandah!

November 1. Night walking with the minister.

Alison showed off to the minister her green-collared sifaka troop. Almost all the reserve sifaka are color-coded for long-term study of individuals. Odd how little-headed sifaka look with collars. Alison says your fingers would meet round their necks. Their heads, so detached, are too small for their bodies. The collars are just nylon dog collars with a buckle, and a tag with a number attached.

Alison, incidentally, was wearing a rather daring sun dress last night, a shirt and *lamba* skirt to arrive in her camp, and now jeans with the same shirt to professionally trot round the forest. Alison hasn't, I think, told the minister about her sifaka age and tooth-wear charts, and the weight charts that suggest sifaka females die young or anyway middle-aged—the surplus males are old (big, horny-toothed ones!). Oh, how I wish we had real long-term data from Berenty!

No hope of supper. Since it seemed to be dark we set out night-walking—a great troop with a torch like a car headlight, trying to spot lemurs by shape instead of using a headlamp for eye-shine. I had to trot ahead and show them how to hold torches. Then the hordes came back, and the minister asked me about *Lepilemur* mating season. He really wanted me to repeat the heresy that lemurs are female-dominant, to shock his troops. I told them Jay Russell's party piece about lepi not urinating in July. Awed interest in the peculiarities of primatologists.

But one does forget what a thrill it is to go night-walking and come out safe—above all for Malagasy who have never dared enter a forest at night. Adrenaline, as Renée said later, is also a drug—night-walking for them is adrenaline and wonder, the edge of fear.

We spotted eyes way down a cut transect tagged with white rings on the trees. We went in after it—way way in, the shine going and going, sometimes one, sometimes two. Someone said, 'Are you sure we won't get lost?' But we kept on—it was straight down the transect path. It became obvious that anything which shone that far in the night had exceptional eyes indeed, like Jules Verne's' Nautilus.

So we weren't too surprised to identify the creature as Roland Albignac lemur-watching with headlamp and torch. Roland said we must be disappointed not to have discovered the eyes belonged to a giant lemur that would give us instant membership in the Académie Malgache, and I said he'd look fine stuffed.

November 2. Bezà Mahafaly is official!

Alison Richard's great day! She danced and rushed and sparkled and organized and was feted, in blue sundress and vertical curls. And President Ravelojaona's great day, and Pothin's, and Andriamampianina's and Bob Sussman's.

A pavilion for speeches—great banner with RATSIRAKA in red across, Malagasy flags at the corners. Two hundred villagers grouped on the ground in a wide circle, wherever trees and houses give a spot of shade. Orations, the meat and drink of Madagascar. A hard-line socialist speech by the politician from Betioky town.

Speeches by the mayor of Ankazombalala and a secretary general of the Ministry of Higher Education. At last the *pièce de résistance*—the minister's speech. The immortal and familiar phrase: *Tsy misy ala—Tsy misy rano—Tsy misy vary!* (If there is no forest, there is no water, there is no rice.) He came through in the end—chubby face gleaming with earnestness and light reflected up off the hot sand outside, and a background of the translucent red, white and green blowing banners.

At last he announced that Madagascar has twenty-two Special Reserves, and Bezà is to be the twenty-third! The others had thought that was coming, but not been quite sure until he actually said it in public. Andriamampianina said afterwards he had tears in his eyes when it came true. The first new reserve in twenty years.

So we ceremonially trooped to the Reserve entrance, where they'd put up the new blue sign. Nicola Blay, the World Bank rep's wife in a Bezà T-shirt, clipped the red, white and green ribbon. The minister ceremoniously shook up a bottle of champagne (The cork broke, and it had to be pried out with the scissors, but that's standard for Madagascar.) After spilling a little for the ancestors, the nearest people passed round and drank the rest. Alison R. and Bob Sussman and Pothin Rakotomanga and Joseph Andriamampianina and Gilbert Ravelojaona stood in front of the blue sign that declares this to be *Ala Tahiry Mandrakizay*—'Forest Reserve Forever,' Alison with the Malagasy ribbon from the opening draped about her neck, and all of them grinning with absolute joy.[5]

5. The formal decree of the Reserve did not come out until 1986, the next year.

❦

That day ushered in the next phase of the Bezà partnership. Now there was funding from USAID to address the urgent priorities. First to fix the road out to Betioky so that villagers could take produce to market. That worked. The second initiative, an ambitious canal to bring water for irrigated rice fields, did not work. The universities and their new partner, WWF, refocused on small-scale projects carried out by villagers themselves. Some were successful; some were not.

Worse, there was major dispute among the so-called 'community' about who should represent them, lead and profit by the new projects. It took years of highly iterated meetings to tease out what people really thought, and the election of a new mayor to resolve the impasse, not until 2003.

Meanwhile, an Environmental Monitoring Team took shape: local people who knew the forest and trained as scientific assistants. They have done good science through the years. In retrospect these men and women became the key to village relations as well. The Liz Claiborne–Art Ortenberg Foundation has maintained its support for this core group and for the Reserve itself through all the ups and downs of political change. Alison writes:

> My favorite monitoring team moment happened when a young postdoc from Harvard was visiting Bezà. She was an entomologist. She came back from the forest the first day, full of triumph, bearing an insect which she believed to be a new species. She asked Edidy (surname Ellis) if he'd ever seen it before and if by any chance he might know the Malagasy name for it. He gave her its Latin genus and species, and explained that it was common. He told me this story afterwards, adding: 'and what I was really proud of was that I didn't laugh; I just told her politely.' Did his science or outreach help relations, I wonder, or just put an uppity Harvard postdoc in her place? Anyway, the whole

> team is unfailingly supportive, helpful and polite to visiting scientists.[6]

Finally, or finally so far, the initial Reserve was in two parcels: 80 hectares of gallery forest, 520 hectares of spiny forest, unconnected to each other. In 2008, after much local consultation, it was expanded to a 4,600-hectare contiguous area: at last a viable size.

There continue to be huge problems, especially the skyrocketing cattle rustling which means most families have lost their principal wealth. In collaboration with the School of Agronomy, Dr. Anne Axel[7] began a livestock research program at Bezà Mahafaly in 2013, to study seasonal grazing patterns. Working closely with three herders, Axel attached GPS monitors to the horns of 100 head of cattle. After three months, 36 cattle remained in the study. Armed men stole the rest. The immediate consequences have been accelerated deterioration of the regional economy, collapse of tourism, and increasing resort to banditry of all sorts by cattle thieves who have run out of cattle to steal.

On the other hand, reports by the monitoring team confirmed that wildlife remains in good health. Giant coua (*Coua gigas*) strut the forest floor defending their territories, still present in numbers first recorded twenty years ago; their abundance signals that these endangered endemic birds and their habitat have been well protected, for they are hunted elsewhere, and considered indicators of undisturbed forest… Of the 212 individual radiated tortoises (*Astrochelys radiata*) marked and measured since 2004, all but 20 have been observed at least once, and often several times. Just 3 have been found dead, from what appeared to be natural causes. This is a real conservation achievement, when it is estimated that

6. Alison Richard, email to Alison Jolly, January 6, 2014.
7. Dr Anne Axel, Marshall University, West Virginia.

around 518,000 radiated tortoises were collected and exported from Madagascar last year alone.[8] The sifaka and ring-tailed lemurs are flourishing too. Cyclone Haruna swept through the region in February 2013, but 34 of the 57 sifaka born last summer are sailing through the trees this August—the largest surviving birth cohort on record.[9]

The trust between the universities and the elements of the community seems to be well founded, after thirty-seven years of coexistence. The main link now is Joelisoa Ratsirarson, son of the Ratsirarson who traveled to St. Catherine's and then became secretary general of Water and Forests. Joelisoa himself has had a dizzy career, rising to become the chief of staff for President Ravalomanana, until the 2009 military coup, and now, as the vice president of the University, an ESSA professor with the interests of Bezà Mahafaly at heart.

Since 2012 there is a whole new venture: Salt of the Earth. The village women have long had a local market for salt. First they scrape it off the ground from seasonally humid sites. Then they carry it 2 kilometers in baskets on their heads to a processing site nearer the village and safer from thieves. Then they mix it with water and filter it through a reed sieve into a hollowed-out tree trunk. They precipitate it by boiling in a pan of corrugated iron, using improved stoves that take relatively little wood, and finally sell it at derisory prices to merchants from Toliary. The salt is prized throughout Madagascar for its healthy properties: higher in potassium than ordinary salt. Alison R. and Joelisoa have taken this in hand: soon you may be able to buy Salt of the Earth in the most exclusive of gourmet boutiques, preferably at derisorily high prices. Other gourmet salts sell from $20 up to $875 per kilo! She writes: 'Looking around at everyone assembled

8. Martin Nicoll, pers. comm.

9. J. Ratsirarson and A.F. Richard, *Interim Report on Bezà Mahafaly to the Liz Claiborne Art Ortenberg Foundation, Antananarivo* (2013).

for a big *kabary* (meeting) in the village of Bejio (one of the two salt-making villages), the Fokontany president turned to me and said 'I have eighty children under the age of 10 in this village. What will become of them when they grow up? This initiative is the first and only hope I have of finding something for them to do.' Alison does have organizational ability: as provost of Yale and vice chancellor of Cambridge she triumphed over quarrelsome university dons. Will she be able to link the quarrelsome women of Bezà with their most hopeful source of income?[10]

Alison and Joelisoa draw five conclusions, of which the first three apply in their own way to Berenty as well:

> (i) the importance of relationships and trust, and the length of time it takes to build both; (ii) the inherent fragility of community-based collaborations, which depend heavily on particular individuals and the pressures on people's lives; (iii) the importance of sustained financial inputs and challenge of diversifying these inputs; (iv) the need for mechanisms to distribute costs and benefits that are accepted as fair, and for methods to track that distribution; and (v) the central roles of improvisation and opportunism in the face of high levels of uncertainty, and the unanticipated key role played by a village-based environmental monitoring team.[11]

Berenty, and above all Bezà, may give the impression that only intensive, small-scale reserves have a hope of long-term success. So far this may be true. There is in the south a huge National Park, some 430 square km, even larger than Ranomafana. It is embedded in a mosaic of protected areas:

---

10. Ratsirarson and Richard, *Interim Report on Bezà Mahafaly*; Alison Richard, email to Alison Jolly, January 6, 2014..

11. A.F. Richard and J. Ratsirarson, 'Partnership in practice: making conservation work at Bezà Mahafaly, southwest Madagascar' (2013).

in all a 22,000 km^2 zone of formal protection. This is Tsimanampetsotsa, in the driest corner of all in Madagascar, with only about 300–500 mm of rain a year. A vast limestone plateau ends in bluffs where the ring-tailed lemurs sleep in limestone caves to keep cool, and because the short, spiky forest offers even less protection from predators than a rocky cliff. Blind cave fish live beneath in underground springs. Below the plateau lies a 15-km-long soda lake, beloved of flocks of pink, greater and dwarf flamingos. The lake itself may look like an arm of the sea, but the outlet loses itself in sand. Tsimanampetsotsa means 'It has no whales.'[12]

People do live here. They farm on the dry plateau, amid the spiny forest. They farm on the sand dunes seaward of the lake. As always, new farmland is cut from the remaining forests. They are acutely aware that times are changing. Rain which used to fall from November now falls mainly in January, if they are lucky.[13] Their herds diminish, through penury and cattle rustling, so more and more they depend on the fields. A rapid socio-economic survey found a highly mobile set of strategies to cope. Traditionally, cattle were moved from lowlands to plateau seasonally to find forage. This continues, but now it is men who move, seasonally or for longer, seeking jobs to supplement the family food supply.

For them, as for the rest of the south, zebu are the only true mark of status. A family starts by raising chickens, saves enough to buy goats, and moves on, if lucky, to one or more zebu. This gives status for funerals and marriages: the living bank account. However, if the zebu are stolen or simply die the bank account and the status crash together. One alternative investment, especially among the people who live below

12. M.L. Sauther, F. Cuozzo et al. 'Limestone cliff-face and cave use by wild ring-tailed lemurs (*Lemur catta*) in southwestern Madagascar' (2013).

13. R.Y. Ratovomanana, C. Rajeriarison et al., 'Phenology of different vegetation types in Tsimanampetsotsa National Park, south-western Madagascar' (2011).

the plateau, is polygamy. An extra wife can farm even more land as well as bearing more children, to the family's delight and pride.[14]

Much of this area is now National Park, with its layers of permitted traditional usage. However, no permits can keep out the zebu hidden in the park away from rustlers.

Jörg Ganzhorn, who has supervised students there since 2000, writes:

> Tsim might be an example of what large parts of Madagascar might experience in the future: too many people for the carrying capacity of the land. The rainfall pattern has changed. Instead of 'reliable' rain somewhere between December and March, the rains now fall several weeks later and we now have rain in August. In 2013 we also had a locust outbreak. One of our project leaders just returned and reports that some people eat twice per week. Several aid organizations are in the region, but logistics are a nightmare and they can not cover the whole region. People rely on the resources provided by the national park. In addition to the usual services, it now serves also as barrier against cattle thieves and a place to hide the cattle. As a result large parts look like the clean forest of medieval Europe when forests were used for pasture and nothing was left on the ground.[15]

Jörg has a radical suggestion. One of the iconic endangered animals of the south is the radiated tortoise. Organized poaching of hundreds of thousands of the endangered tortoises for soup or elite pets for Asia, or even just bushmeat for local consumption, may wipe out what was once a population of millions within the next twenty years.[16]

---

14. SuLaMa, *Diagnostic participatif de la gestion des ressources naturelles sur le plateau Mahafaly* (2011).

15. Jörg Ganzhorn, email to Alison Jolly, January 4, 2014.

16. H. Randriamahazo, 'Radiated tortoises and the fading taboo' (2011); R.C. Walker and T.H. Rafeliarisoa, 'The precarious conservation status of the critically endangered

(I admit to once trying radiated tortoise stew back in the 1960s. It tasted to me like boiled gray art erasers with onion sauce. Perhaps it needs a gourmet recipe.)

Jörg's idea:

> I would sell tortoises. On the European market 1 tortoise sells for the equivalent of 10 zebu or the yearly salary of a teacher at a primary school or the salary of a nurse for 8 months.

This would bring in real income, but like the Salt of the Earth initiative, it would take savvy marketing and control on the export side, as well as changing the current, unenforceable law. Then there are all the problems of fairly distributing returns to the park and people concerned.

Still, Jörg ends on a hopeful note:

> My grand-grand-grandfather was a shepherd in southern Germany in an area about as fertile as the Mahafaly plateau. He walked with his lambs once per year from southern Germany to Paris to sell them; again, about the distance between Tsimanampetsotsa and Tana. Now the southern region is the richest part of Germany. If we can do it, Madagascar can do it.[17]

Madagascar spider tortoise (*Pyxis arachnoides*) and radiated tortoise (*Astrochelys radiata*): what we now know through three years of field operations' (2011).

17. Jörg Ganzhorn, email to Alison Jolly, January 4, 2014.

TWENTY-TWO

Climate change

More drastically, the future prognosis is not more of the same, but worse. More and more people turn to food aid. There is also agricultural aid, like efforts to reintroduce sorghum as a drought-resistant crop, and many small irrigation projects to use water wisely—when there is water. So far, no one has turned the south into a place sustainably self-sufficient through bad years.

Madagascar is among the countries predicted to be most threatened by climate change. There are cyclones and floods and droughts. And swarms of locusts. Part IV opened by saying that Madagascar has more unpredictable weather than most of the world. The evidence comes from recent weather data, but even more tellingly from the adaptations of so much of the fauna with bet-hedging life histories so that they may survive catastrophic years. How can people and animals survive in the future?[1]

Climate predictions, in Madagascar as elsewhere, are subject to many uncertainties. There is the problem of background climate variability, of estimating rich- and poor-country worldwide emissions, and also the conflicts between

1. Maplecroft Global Climate Change Vulnerability Index, http://maplecroft.com/themes/cc.

different gross climate models and the problems of 'down-scaling' the global models to predict what is happening in any one place. From the historical record we know there has already been warming in Madagascar of about 0.2 degrees since the 1950s, and a decrease in rainfall, with later onset of the wet season when crops are grown.[2] Temperature is expected to increase by up to 3 degrees in the next forty years, with the highest rise in the south—which is already sweltering at 30–40 degrees during summer.[3]

The crucial question of rainfall is disputed. Some models suggest that rainfall will increase over Madagascar in most months, and during the wet summer season even in the far south. This would be very good news for southern farmers. Others suggest much longer and more intense droughts. However, winter rainfall in the south is predicted to decrease. This is bad news for drinking and washing water, but people adapt to that. It may be more problematic for the spiny forest. Madagascar's spiny forest is an almost unique ecosystem with drought-adapted plants that form a thick, thorny woodland. It may have adapted to having a small amount of rain even during the driest season. If this is true, drier winters will force the plants toward becoming not spiny forest, but spiny desert.[4]

The most drastic weather patterns, though, are cyclones. Whereas projections 'indicate decreases in the frequency of cyclones during the early part of the main season, their intensity, associated winds and destructive power are all suggested to increase as we progress towards the end of the century.'[5] Madagascar suffers at least one cyclone in most years: in 2007 there were six. On average 250,000 persons are

2. M. Tadross, L. Randriamarolaza et al., *Climate Change in Madagascar; Recent Past and Future* (2008).
3. World Bank, *Madagascar Country Environmental Analysis* (2013).
4. Ibid.; P.J. Grubb, 'Interpreting some outstanding features of the flora and vegetation of Madagascar' (2003).
5. Tadross, Randriamarolaza et al., *Climate Change in Madagascar*, p. 8.

affected, and some $50 million worth of damage is caused by each cyclone event. They obliterate locally built houses and, even worse, can destroy a year's crops. For people depending on subsistence agriculture, this means no income and no food for a whole year unless they can find help. Cyclones are not expected to increase in frequency, but all models predict more intense cyclones—New Yorkers can imagine what that means!

No, New Yorkers cannot imagine, because their year's food supply was not wiped out. Cyclone Haruna, of February 22–25, 2013, displaced at least 10,000 people in the south-west regions of Tulear and Morombe. A few people drowned; the rest are putting their lives back together. In New York that means massive state aid, reconstruction, pumping out the subways. Lingering annoyance as electricity is restored: most places up in a week, but, say, an elevator to the High Line not back in services for six months. In Tulear region, on the other hand, this means no official help to rebuild one's house, and certainly not to replace the crop drowned in the field. Oh well, you can grow some more next year—in the meantime, just cope.

In 2013–14 the crop may be eaten up by locusts. Cyclone Haruna provided ideal conditions for the coming swarm of ravenous insects. Their numbers were already building up last year: FAO warned the government and the World Bank of an urgent need as early as August 2012. Nothing was done, so now FAO says that pesticide spraying from helicopters should have started in September 2013—and that it will take some $41.5 million to contain the outbreaks over the next three years. Some two-thirds of Madagascar's food supply is under threat. The Bank has offered a $10 million loan for now...

Have you ever seen a locust outbreak? Jeffrey Katzenberg and Stephen Spielberg take note: this is the ultimate alien invasion. The sky glitters. Millions upon millions of sparkles of light reflect the tropical sun. They descend. The sparkles turn into dark spots before your eyes like the last stages of dizziness as you fall into coma, but there are millions of them and the sky roars. And then, on the ground, the seething mass of yellow bodies, tiger-striped with brown, the shield-shaped thorax hunching over each chewing head like the armor of a surrealist robot. Millions upon millions, which may have flown 100 or 200 kilometers in a day on those glittering wings. And after? A brief fry-up that evening, then your family's food for that year obliterated.

What about other fauna and flora? Reptile species are already moving uphill—but what happens when there is no more mountain above the peak?[6] North–south corridors would be ideal for the dispersal of endangered species, but Madagascar's forests are already highly fragmented—planting new forest corridors would need huge funding and buy-in from local people. Lee Hannah and a raft of colleagues attempted to model predicted changes in forest cover, and to estimate the cost of maintaining connectivity as of 2008: about 0.8 billion dollars. Since conservation funding for protected areas is only around $9 million, we are very far from what is needed—even assuming government and people are interested in doing it.[7]

The bottom line is that change is coming. It is all very romantic to dance to the cowhide drums and home-made wooden violins at a big funeral, and to play with the hordes of bright-eyed children. When the extended family, and the herds, and emigration to towns are not enough, Madagascar needs a radical change in the possibility of feeding its people.

6. C.J. Raxworthy, R.G. Pearson et al. 'Extinction vulnerability of tropical montane endemism from warming and upslope displacement: a preliminary appraisal from the highest massif in Madagascar' (2008).

7. L. Hannah, R. Dave et al. 'Climate change adaptation for conservation in Madagascar' (2008); Raxworthy, Pearson et al. (2008). 'Extinction vulnerability of tropical montane endemism from warming and upslope displacement.'

January 20, 2014.

As of this weekend there were two major announcements in the press. On Friday the 17th, the final results of the Malagasy elections. I will come to those in the final chapter. But far more important, on Sunday an article by Wenju Cai and his many collaborators predicted increased frequency of extreme El Niño events with global warming. If this is right, droughts in the USA, landslides in South America, bush fires in Australia and Indonesia will cause ever more frequent extreme destruction. Not to mention hunger in Madagascar...[8]

8. W. Cai, S. Borface et al. 'Increasing frequency of extreme El Niño events due to greenhouse warming' (2014).

PART V

Money

The final part takes us to the present day. To recapitulate, we began in Part I with the joy and tragedies of life in 'traditional' villages. The villages in fact were already impacted by much wider influences—but village structure remains the basis of rural Malagasy life. Part II was the politics of conservation in the 1980s, with the Western ideal of saving biodiversity and the mutual incomprehension between conservationists and Malagasy politicians. Part III was a case study of the difficulty of putting the ideals into practice at Ranomafana: the triumphs of research on rainforest biodiversity set against the effort to 'develop' peasants who have lost the land bequeathed by their ancestors and the livelihoods given by that land. Part IV moved to the semi-arid spiny forest, with the impact of drought on both lemurs and people, and the conclusion that traditional ways of life do not cope with rising population and a shrinking resource base. Both rainforest and spiny forest livelihoods must change quite fundamentally even now, let alone with the more severe weather looming through climate change.

In this final part, I hazard some opinions about what is being done—or not done—to improve life in Madagascar. Chapter 23 briefly describes political changes from 2002 to 2013: two presidents with starkly different approaches to governance. The 2009–13 rosewood massacre looted the National Parks for short-term gain. Both regimes gave in to land grabs for Asian food. Chapter 24 turns to economic changes. Is mining a promise of solid finance and indeed environmental ethics? Or inevitably the curse of corruption? The REDD initiatives, Reducing Emissions from Deforestation and Forest Degredation, actually pay to preserve forest carbon stocks. Is REDD the best hope ever of paying Malagasy the value of their forests? Or just another kind of land grab?

And a coda of 2013–14. Election run-ups which sometimes descended into farce—if only it didn't matter! And yet another primatological congress in Ranomafana in August—a celebration of Madagascar's glorious biodiversity thirty years after the 1985 conference which opened this book, with a whole new generation of Malagasy field scientists.

That 1985 conference embodied the hope for a win–win strategy for both environment and development. It has yet to happen. But it still might do so. I wish I could stay around to see it.

TWENTY-THREE

Durban Vision; rosewood massacre

I've seen all the presidents of Madagascar since independence, though only met a couple of them. Presidents in Madagascar have far more power than in countries which pride themselves on checks and balances—presidents and their circle of henchpersons tell the country what to do.

Of course, many don't listen. Villagers try to keep their heads down. Their experience of officialdom is foresters with power over land, petty clerks on the take, and literate people who ask them to sign incomprehensible papers which usually turn out to mean they have relinquished their rights. Among bureaucrats there is still a core of dedicated people who would very much like to run an honest and efficient civil service. Businessmen of course seek profit, but profit may be had either by weaseling round the system or by straightforward investments that yield jobs and growth as well as profit. But, still, it is the president who holds the financial reins, and whose policies set the country's course internationally.

They have not done a very good job. Madagascar has declined in real per capita income ever since 1970. Mireille Razafindrakoto and her collaborators at the *Institut de Recherches sur le Développement* illustrate this starkly (overleaf). The left-hand graph shows the decline in per capita purchasing power compared to other African states. The right-hand graph

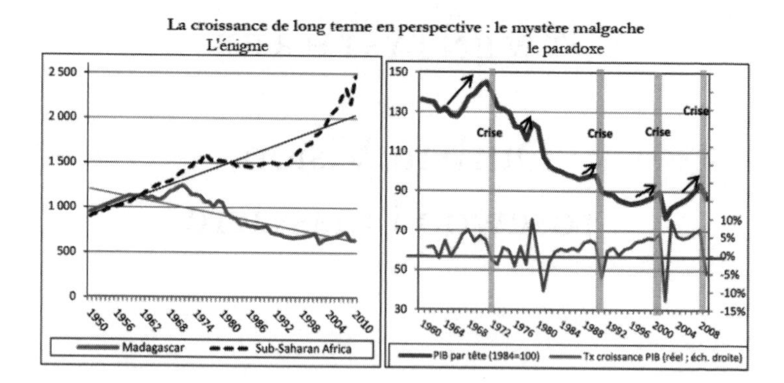

shows the modest rises under each regime and the precipitous falls, with very slow recovery, at each regime change.[1] As the economy grows, 'so does the frustration of the excluded groups... Add to the mix a dash of weak institutions and a pinch of opportunistic disgruntled politicians, and you have all the ingredients of the 'Madagascar Cycle.'[2]

Two questions. Why have the Malagasy put up with this for so long? And why has no president managed to change the trajectory?

Madagascar has a highly hierarchical society. Long before colonization the ruling Merina recognized some twenty-seven social castes, ranging from the highest nobility down to slaves at the bottom. French colonists moved in at the top, but without erasing the memories of who used to be whom. The Merina tribe also conquered many of the coastal tribes. 'Ethnicity,' or, to be frank, racial discrimination between more Indonesian plateau people and more African coastal

---

1. M. Razafindrakoto, F. Roubaud et al., *Institution, gouvernance et croissance de long terme à Madagascar: l'enigme et le paradox* (2013). This report offers an extensive analysis with interviews and statistics to explain why people accept the stagnation of the economy under successive regimes. Don't worry about the graph axes: the first is in constant 1984 Malagasy francs, the second in purchasing power of constant 1990 (Geary-Khamis) dollars. The point is the direction of change.
2. S. Andriamananjara and A. Sy (2013). 'It's time to break the 'Madagascar Cycle.'

people, is still strong. It matters how light or dark your skin is, how straight or kinky your hair. But underlying all these gradations is a kind of inertia or reticence or respect toward one's superiors.

Malagasy have an especially strong tendency to accept those in power. They may see top politicians as ruling by right—to be *raiamandreny*—father-and-mother—a traditional term of respect for wisdom and semi-sacred power and fertility. This gives society an apparent peacefulness: a restraint which means that even when presidents have changed the country has not erupted into real civil war. It is true that violence can break through in mob lynching of a thief or mob looting of Indian shops. However, after the eruption, people regain an extraordinary surface of quiet etiquette toward each other. This is not just fear; it is tradition.[3]

The next question, why no regime has stopped the economic decline, is complicated. External pressures, from commodity price falls to imposed structural adjustment to the burden of debt, have not helped. But a part of the answer comes with Donald Stone's remark, back in Ranomafana, that hope is a terrible risk for those in power. As conditions improve slightly, but never fast enough, people who have been left out take matters into their own hands and back some new politician who promises change.

Enter Marc Ravalomanana.

In early 2002, Marc Ravalomanana, the mayor of Antananarivo, won the presidential election. Didier Ratsiraka, who had been president for most of the years since 1975, did not accept the results. Six months of unrest followed, the 'crisis' with massive pro-Ravalomanana demonstrations in

3. C. Alexandre, *Violences Malgaches* (2007). An attempt by a French priest to understand Malagasy peacefulness and sporadic outbursts, as well as why circumlocutions in meetings baffle expatriates—so often the speaker is unwilling to commit himself until hearing others.

Tana, led by Lutheran exorcists to clear away the 'devils.' Demonstrations in the mornings, actually tidying up the Avenue de l'Indépendance by noon, after which people went to lunch and back to work. Meanwhile Ratsiraka's militia dynamited the seven bridges to Tana, starving the capitol of petrol and other goods. Porters carried jerrycans of petrol across the river beds, for a price, sometimes smoking all the while. There was a real battle between army factions, in which 14 people were killed: 8 on one side, 5 on the other, and a woman who walked by at the wrong moment. There was no blood-bath, which one might expect in actual African countries. In the six months a total of some 100–150 people were killed, many to settle old scores or to loot their property. In the end, Ratsiraka summoned some old-time French mercenaries to help him. France had its eye on these veterans of other mercenary wars, so their plane was turned around halfway to Madagascar. President Ratsiraka promptly fled with family and friends.

Perhaps more people died of flu than of violence. At the same time as the 'crisis,' a strain of flu was killing villagers through much of the eastern forest. It was a banal strain: people in the west with a flu jab were already immunized. However, in a population severely malnourished, most of whom wouldn't pay for medicine until at the point of death, it swept away the old, the very young, the infirm.

My colleague Hanta was in Tana during the 'crisis.' Her sons in their scout troops and aikido clubs volunteered to help patrol the capital. Richard and I did not arrive until Ravalomanana was firmly installed as president. The president was just about to fly to the UN, where he pledged to support the environment with an actual percentage of government revenue.

The extraordinary ex-ambassador Léon Rajaobelina put him up to it, in his post as director of Conservation

International Madagascar. In this new capacity Léon could be an *éminence grise* both within and outside the country. It seemed that conservation was at last firmly on the national agenda.

Letter to family, Oct. 12, 2002.

Our 39th wedding anniversary!

This place is weird. There are now computers. There is even a phone book. Also even more child beggars on the streets, and grimy poverty: I think visibly worse than ever, though maybe I just know it is. Sample: the car slows in the endless traffic jam, and is surrounded by two different men selling car accessories: bumpy massage back-mats for drivers, a leopard-skin steering-wheel cover. They will never own a car or a steering wheel. Also two children: girl, maybe 7, and boy in an orange T-shirt with literally more holes than shirt, maybe 5. Bright-eyed, but of course they are stunted so are older than they look. Also a hand reaching up from below: a polio-crippled beggar on a rolling cart made of a plank on little wheels. The car lurches forward 10 feet and then we have the next lot of hopefuls.

The last time that I was in the Presidential Palace I was with a Canadian tourist, who was frogmarched into the guard-house for daring to snap photographs...

This time we were escorted in the front, between the curly iron gates, up to the ever-so-French-provincial facade of pink brick and white trim, three storeys of colonial pomposity. Set, indeed, in a lovely garden of purple-flowered jacaranda trees and a wide view of the hill town of Antananarivo. The anteroom where we waited was two storeys tall, arched French windows onto the garden, filmy white curtains and sunshine-yellow ones blowing in the breeze. We sat round a great polished table waiting our turn. The president's schedule was a bit behind: a gang of portly oil men came in and sat down at the other end of the table to wait their turn after ours.

Ushered in: just one lackey. No obvious bodyguards. Marc Ravalomanana: a slight, erect, handsome man sitting at a long table in front of an inlay-on-black mural of plateau scenery, complete with women bending to plant rice, and oxen drawing carts. The president rose to greet us, shaking hands all round. Does he look less handsome than his pictures? A few more lines round the eyes. After all, he has taken the decisions for six months, first to call out the entire population in Gandhi-like resistance demonstrations, and then for the lightning-fast movement of troops to take over the towns whose governors still supported the old dictator. Also, he has just flown in from the World Summit on Sustainable Development in Johannesburg, where he made his maiden international speech as a head of state. Also, he's leaving this afternoon for the General Assembly in New York. A few more wrinkles are understandable.

Richard, Gouri Ghosh and Eirah Gorre-Dale of Unicef made their pitch about hygiene and sanitation and the WASH[4] campaign. He was polite, but not well briefed or enthusiastic. Maybe if they had a glitzier promise—to deal with the banal strain of flu that has claimed hundreds of lives of the very poor, or to deal with the economy set to deliver a minus 10 percent growth rate this year after the six months' siege of the capital—then he'd have had more to say.

Maybe I got the happiest reaction. At the end, I said how much I admired his Johannesburg speech, especially its preamble that Madagascar is one of the countries with the greatest stake in the environment because practically all its wild fauna and flora are unique. I said that according to me

4. Water, Sanitation and Hygiene campaign, supported by Unicef, Water Aid and others.

and all my colleagues, Madagascar is one of the richest countries in the world.

'I love fauna and flora, and we are indeed rich!' he proclaimed, rising. 'You see, even my office has orchids!' It did: three pots with fronds of pink and white Madagascar orchids, each flower spray taller than a man. So we wound up taking his picture with his orchids, as well as with ourselves. And then we took our leave, while he went on with the affairs of state.

And the rather more substantial interview with the oil men…

A year later, in September 2003 at the World Parks Congress in Durban, Ravalomanana announced the 'Durban Vision.' Madagascar would triple the extent of its protected forest areas up to 6 million hectares. The new areas would be community owned, and more or less community managed. This reversed more than a century of colonial and pre-colonial fiat that all forest land belonged to the state.

A consortium of advisors implemented the 'Durban Vision,' led of course by Léon and Conservation International. Madagascar then designated 2 million hectares of new forest reserve areas in each of three successive years.

Of course, with all this haste, there were many cases of misallocation. Often the people who came to the consultations on the ground did not actually represent their local communities. Sometimes they did not include sacred forests in the scheme, thinking these were already protected since time immemorial, thus missing out the most important areas to benefit. Above all, although the forests were to be under community management, few at any level were clear quite what this would mean. Still, it was a huge lurch forward in the idea that people themselves would and could manage their own landscapes. Even a paper designation of a reserved forest goes some way toward its protection.[5]

Over the next years, Ravalomanana implemented many of his other election promises. Hundreds of new schools were built, teachers paid. Every child received a backpack with a new notebook, pencils and rulers, of which they were inordinately proud. Clinics opened. A free-trade zone in Antananarivo mushroomed into a thriving textile center. A nationwide election in 2006 confirmed Ravalomanana's mandate as president (but the Malagasy nearly always re-elect a sitting president). To the fury of France, he announced that English should be taught in all schools, with the goal of Madagascar, like Mauritius, becoming a trilingual country able to operate in the Internet-connected world.

It wasn't squeaky clean. The Ravalomanana enterprises prospered greatly, from computer imports to dairy products. But the self-made millionaire went on coining money from long-term, productive investments in the country, expecting that his businesses and his presidency were in it for the long haul.

Then Ravalomanana made a big mistake. He allowed foreigners to own land. A huge land grab in the west of some 1.3 million hectares was leased for ninety-nine years to the Daewoo company of Korea to grow maize and oil palm. Domestic and foreign press screamed about that one. It wasn't mostly arable, because the small patches of arable land in Madagascar are bisected by grassy hill-pastures, but it was hard to believe that Daewoo's promises of employment and investment would in any way compensate local people who had lost their farms and their grazing rights—let alone address the semi-sacred status of ancestral land. Still, that

5. C. Corson, 'Territorialization, enclosure and non-state influence in struggles over Madagascar's forests' (2011); C. Corson, 'From rhetoric to practice: how high-profile politics impeded community consultation in Madagascar's new protected areas' (2012).

was in the far west. Ravalomanana, who had made his first fortune from his family's yogurt business, was all for improving cattle and dairy yields. He very obviously paved over rice fields on the road from the capital to the airport for a factory to produce high-quality animal chow. That was an immediate and obvious insult: turning productive rice paddies into an animal food factory was visibly disgusting.

His even bigger mistake: he built schools and clinics, as he later wryly remarked, instead of raising army salaries. In March 2009 the mayor of Antananarivo, Andry Rajaoelina, staged a coup. Mobs in one night burned down several of the Ravalomanana factories and his party-run radio station. There was some loss of life—mainly among the rioters caught in the flames while looting the Ravalomanana-owned supermarket. A faction of the army supported Andry.

When I first saw Andry's posters in his mayoral campaign, I thought they were advertising a pop star. Not so far wrong: 30-year-old Andry had been a disc jockey. Someone backed and bankrolled the young man, of course, but I am not repeating here the rumors naming obvious suspects—mostly French companies encouraged by a French government. Anyway, Andry wasn't much younger than President Ratsiraka back in 1975.

Then Ravalomanana, or his adjutants, did just what Ratsiraka had done in 1991: they fired on a crowd. At a demonstration in front of the Presidential Palace some seventy protesters were killed. Among them was my friend Hanta's eldest son on his first job as a journalist standing out in front of the crowd holding a microphone. Hanta, her husband Niry, and all of us who knew them were stunned with grief, as were so many other families.

Ravalomanana fled to South Africa.

The new government was flagrantly illegal. Foreign donors promptly froze all non-humanitarian aid. Much of government expenditure and all infrastructure improvement were wiped out: foreign aid and loans had been 40 percent of the government budget. Educational enrollments plummeted. Health clinics closed. Businesses in the free-trade zone packed up and moved to other countries.

❦

The 'transitional' government plundered its honey-pot: the rosewood of the eastern forest reserves.

It is estimated that rosewood worth a quarter of a billion dollars left Madagascar for China in 2009 alone. And as much again in 2010. The quantities have tailed off since—to some extent. The shipments are and always have been illegal. The proceeds were obviously not much to Madagascar: Chinese importers are not so kind. Nor did much go to people who actually risked their lives felling giant trees on slippery slopes: their wages were under $2 a day. Even that, though, was a wage for an unemployed laborer, so there was no dearth of candidates for the jobs. The main proceeds, instead, were split between the timber barons of the east coast and the people at top levels of government who took their cut.

Rosewood is dense, beautiful hardwood. A cut rosewood trunk does not float. To slide a cut trunk to the nearest stream you have to hack a slipway out of the mountainside. Then you cut three to five other trees of lighter consistency. You tie them to the rosewood to raft it down the river. It is piled onto trucks at the landing stage—diesel dinosaurs like the one Eleanor Sterling and I rode to Ivontaka twenty-five years ago. It is stacked in warehouses, and then loaded onto ships with falsified manifests, and then conveniently overlooked by customs officers. When a cargo is spotted it is the low-level customs people who are fined and jailed, never the bosses.

This didn't pass unnoticed. There are heroes of the struggle, like Erik Patel and his Malagasy colleagues. Erik was a graduate student from Cornell studying silky sifaka, his Ph.D. much delayed by the assault on his site and his animals. Silky sifaka are one of the world's rarest primates, creatures with pink and black spotted faces and an aureole of long white fly-away fur. They live only in the national park of Marojejy: vertical mountain slopes clothed in dense rainforest that used to be full of rosewood trees. Erik, expats and brave Malagasy managed to get articles into the *National Geographic*, under-cover films on the BBC, outrage in the journals of the world. The rosewood massacre has slowed, but not stopped. In the USA Gibson Guitars was sanctioned for using tiny slivers of Madagascar rosewood for fretboards. In my opinion, to use a few bits of rosewood to make fine sculpture or sonorous music should be applauded. Meanwhile no one has stopped the Chinese carving 'Ming dynasty' four-poster beds, each post a whole tree. I saw one bed advertised on the web at a million dollars.

January 19, 2010.

Léon Rajaobelina is off to Washington this week to testify in front of the subcommittee on Africa of the Foreign Affairs Committee of the House of Representatives on the impact of climate change on African countries. Also to lobby the US Congress to lift sanctions on the environment sphere. The rosewood massacre will continue unless the civil service is paid, including the Eaux et Forêts and the National Parks Service and some of the police.

Unicef is trying to step in for the community health centers—hence I joined the three-Land Rover mission to the south. A great deal of aid has been reclassified as humanitarian, via Unicef, but you can't expect other people to provide all the essential services of a country, let alone guard its National Parks.

The international community can only have sanctions against the populace, not against Andry—because he is a French citizen!

It was inconceivable a year ago that Mad. would join the roster of failed states. It hasn't happened yet, but it is now conceivable.

On the other hand, it has not failed yet as badly as some. Björn (on our Unicef mission) made a present to Bruno the Unicef rep of a crisp, new, small-sized banknote from Zimbabwe, value $10 trillion. He regrets having missed the largest version, for $100 trillion, which is as high as they went before switching entirely to US dollars.

But the real prize is revenue from minerals and oil. The rosewood massacre was a mere stopgap. Big mining revenue is still in the future—but arguably the 'transition' of Andry's cronies was sparked by the rush under the previous Ravalomanana government to open concessions for minerals and oil.

Madagascar, which has suffered so much poverty, may now be set for the final insult: the Midas curse of wealth.

TWENTY-FOUR

The new mines

From everything I have said so far, no one should doubt that Madagascar needs money. It needs money wisely distributed among the population. It needs money wisely used for long-term investment. It needs money if it is ever to achieve a sustainable relationship with its environment: soil, water, wood and the jewels of its biodiversity. This chapter argues that in some ways the mines have taken the role that conservationists hoped for in the 1980s: the rich foreigners who can sway government policy. Their goals are different, of course, but before leaping to the conclusion that profit-making capitalism must be bad, consider this.

The life of a mine is thirty years at least, sometimes fifty or sixty. Longer than an NGO five-year grant; much longer than the horizon of a politician who may be out at the next election. Mines have to live with the environment and social structures they create. Mines need huge water supplies and management, an environmental challenge that ramifies into the forests on the surrounding hills and irrigation of the surrounding fields. Besides this, big international mining companies are oddly more vulnerable than little governments. NGOs and shareholders and the Extractive Industries Transparency Initiative (EITI) watch their records and scream if they go wrong—not in some remote landscape, but right at corporate headquarters. The world watches the big mines in a way it rarely cares about impotent Madagascar. A big mining company has its own reasons for ethics: business case reasons. As one World Bank rep remarked to me, 'These people are not choirboys.'

This does not apply to fly-by-night 'juniors.' These are companies which grab an option to open a site, then do so with maximum speed and minimum care. They may do the dirty work, then sell on to larger companies which wait behind, rubbing their clean hands in anticipation. They can also covertly bribe a politician for permits, expecting to make their pile before that politician is ousted. The National Office of the Environment has recently found the courage to roundly condemn one Chinese oil company for its land grabs, its lying to peasants and despoiling of the environment—a rare case which actually touches everyone from the government in power to a brother of George Bush's. Much mining is indeed dirty in every way—but not all mining.[1]

By the time the Marc Ravalomanana regime ended, every part of Madagascar was divided into mining exploration zones: oil, coal, bauxite, gold, sapphires and emeralds, uranium, titanium-rich beach sand. The French geological service identified almost all of these deposits long ago: the riches of Gondwanaland. I can't resist quoting my own oh-so-colonialist letter home from my first impressions:

Jan. 10, 1963. Ambatomika.

Ambatomika, a village founded to mine thorianite.[2]
M. David, chief of the mine, received us, fed us lunch, showed

1. N. Raonimanalina and W. Fitzgibbon, 'Chinese oligarch could face scrutiny in Madagascar oil land grab' (2014).

2. Thorianite: thorium oxide, ThO_2, with other radioactive minerals. It went to the growing French missile and nuclear programs. Wikipedia says thorianite is hard to shield because of high-intensity gamma rays. Drivers for Ambatomika had radiation badges, but perhaps not the women in the *laverie*, who sieved out the 'metal-stone' from crushed rock. One friend from Berenty remembers the *laverie* as a good job, because

us all over. Did I say this is the most primitive part of Madagascar? The Tandroy people live in tiny huts of wood which you cannot stand up in, the men wear loincloths, the women have only just learned to tie their dresses above the bust. In parts of this valley people walk 25 km for a jug of water, which they dig from a dry river bed. Those people mine radioactive minerals at Ambatomika.

M. David showed us everything—chunks of the raw mineral in matrix, maps of how it is found in cloud-like lenses, not continuous layers, the great stepped pit where they have almost finished one lens, the tiny 'Laverie' where thorianite is washed out of the crushed rock. It was terribly impressive to see in this country, where the first word you learn is *mora-mora*—slowly, slowly! Lights over the pit let them work twenty-four hours a day. Bulldozers and dump-trucks operate here, as at home. It is still half and half; one of the fanciest trucks picks up 10-ton bins and carries or dumps them elsewhere. This is handy since the truck can carry one bin while another is being loaded by hand by men with little shovels.

❦

Nowadays, though, it's a different story. Mineral prices soared and demand from China galloped ahead. Not all the concessions are being explored, but dollar signs blinked in everyone's eyes. It is possible the coup of March 2009 had more to do with anticipating the rich pickings of the future, a 'resource curse' in advance, than with local rent-a-mobs, even if the mobs were all too willing to be rented.

Two major mines were already well under way before the coup. In 2005 Rio Tinto made its investment decision in the titanium sands of Fort Dauphin after a twenty-year lead-up

---

you could work at the sieves sitting down. A. Jolly, *Lords and Lemurs: Mad Scientists, Kings with Spears, and the Survival of Diversity in Madagascar* (2004).

of exploration, research and forest regeneration. This would involve destruction of some of the last precious remains of coastal forest. In 2008 Ambatovy Mining SA, a joint venture with Canadian Sherritt, and Asian financing, announced it would build a huge nickel and cobalt open pit at Ambatovy in the rainforest. It needed a 220-km pipeline for slurried ore down to the port of Toamasina and a huge processing plant by the coast. It began shipping ore in 2012. Ambatovy itself, 'The place of iron stone,' is hardly luxuriant: a plateau covered by ferrous mineral crust and straggly trees with very few endemics to that formation. It repeatedly burns: I would not like to be up there in a lightning strike. However, much of the ore body lies under pristine mid-elevation rainforest. The whole area targeted for mining is surrounded on three sides by an even larger area of wonderful forest, which connects in the north-eastern part of the concession, through an as-yet-unprotected corridor, to the main set of protected areas that runs north for 150 km to the reserve of Zahamena.[3] It also lies beside the Torotorofotsy lake and swamp, home of the largest known population of the greater bamboo lemur. Furthermore its pipeline cuts through, or rather under, luxuriant rainforest. Ambatovy Mining agreed to lengthen that line by many kilometers to minimize the environmental damage, and to manage most of their concession that will not be mined for conservation.

We will come back to Rio Tinto, but these two companies have already committed both corporate pledges and millions of dollars in funding for the environment.

Next on the scene came WISCO, the Wuhan Oil and Steel Corporation. In 2010 it paid the Transitional Government $100 million for permits to explore for iron in the Soalala region. The Transitional Government crucially needed a cash

3. P. Lowry, pers. comm.

inflow, starved of foreign exchange after aid was cut off and legitimate exports plummeted. The WISCO concession again lies in an environmentally sensitive area. Its exploration zone covers 430 square km around its future port at Soalala. Soalala lies on Baly Bay, the only known home of plowshare tortoises, the world's rarest tortoise. WISCO originally hoped to export ore within five years, but, with the world slow-down and drop in mineral prices, it may be delaying construction in a way that QMM in Fort Dauphin and Sherritt in Ambatovy are unable to do because they already had committed bulldozers on the ground and ore flowing out.

As for oil, the blocs assigned are even larger. The French company Total and Houston-based Madagascar Oil have huge concessions for tar sands in Bemolanga and Tsimiroro, abutting the world-famous Tsingy de Bemaraha National Park. Environmentalists crowed in 2011 when Madagascar Oil announced it was scaling down its plans for Bemolanga to mere prospecting. The sands have half the concentration of tar as Canadian ones, 5.5 percent versus 11 percent, with correspondingly more cost, oil, water and energy needed to extract the oil. Not to crow too early: Tsimiroro has already started up with 10,000 barrels of oil from tar sand by August 2013, from a pilot steam flood extraction plant and a development plan leading on to full-scale production. It seems that part of the delay was that the Malagasy government wanted a whole $9 million in back-taxes, now negotiated down to $4.5 million—with contracts that allow the oil companies to keep 99 percent of the profits for the next ten years, just 1 percent for the host government.[4]

The social side is even more complex than the environmental side—as well as being fundamental to preserving the environment. Miners have to live with the disgruntled (or

4. G. Gwinnett, 'Madagascar Oil gears up for an exciting six months' (2013); Madagascar Oil, 'Steaming towards First Oil' (2013).

gruntled) populace around them. Disgruntled people simply pile huge stones across a mine access road. They camp beside the road, inviting in television crews. Even if those people are forcibly cleared or bought off in a few days, that means hundreds of thousands of dollars' worth of delay and disruption to the mine, as well as all the negative publicity. Again the mines have a strong incentive to head off revolt by dealing fairly with the local populace.

My own experience has been with the most visibly ethical of mines: Quebec Iron and Titanium Madagascar Minerals, the Fort Dauphin subsidiary of Rio Tinto, which I will shorten to its usual name, QMM.

Let's start with geography. The south-east corner of Madagascar. The narrow coastal strip around Fort Dauphin hemmed by the seacoast and the eastern mountain chain is called *Anosy*, the Island. In many places in Anosy you cannot see the sea, but almost every view includes the mountains. Granite cliffs rise to improbable peaks, their highest slopes still clothed in rainforest. The mountains reach little more than 2,000 meters, indeed 1,000 meters on the side toward the sea, but that is enough to catch the wet wind perpetually blowing in from the Indian Ocean. Young Philippe de Heaulme once described Fort Dauphin as 'The Pisspot of Madagascar,' though in fact it has only about 160 cm of rain a year, much less than further north, and indeed less than the peaks themselves. Inland the mountains create a 'rainfall fault-line.' The semi-arid spiny forest on the western side almost abuts the rainforest above, with around 50 cm of rain in an average year, though no one year is average.

Anosy holds scraps of a still different ecosystem: the coastal forest on white sand. Coastal forest is adapted to resist salt winds and the drying-out effect of roots in white sand, which just drains water away. Most leaves are stiff, wax-coated. After rain the forest glistens as sunlight reflects and refracts from water drops caught on every leaflet. If you want to see such a plant, look at the stephanotis in a bridal bouquet (sometimes called wax flower or Madagascar jasmine): delicate-looking white stars all set to survive being hurled at a bridesmaid.

Most of the coastal strip is now grassland studded with fire-resistant, nutrient-salvaging travellers' palms. Coastal forest is the most threatened forest type of Madagascar. It survives only as tiny patches, each a scientific treasure. The 600 hectares of Petriky, just west of Fort Dauphin, hold twenty-two different plant species unique to that patch alone. Of course they were more widely distributed before people cleared the area around, but now they are just in Petriky.

Andrew Lees died in Petriky. Lees was national campaigns director for Friends of the Earth in Britain. His last campaign was to stop the prospect of mining by QMM. On New Year's Eve 1994, Lees went into Petriky forest alone, telling his taxi driver to come back for him in a couple of hours. He collapsed with heat exhaustion. The taxi driver did say he'd lost his white man, but the local villagers were too frightened to report a corpse in their woods. His body was found after a week, his film camera beside him.

Andrew Lees became an environmental martyr. Protestors from Friends of the Earth marched in front of Rio Tinto's London headquarters in St. James's Square brandishing placards that read 'NO MINING IN MADAGASCAR!' In the south of Madagascar, Rio's vulnerability was clear. Many NGOs, especially WWF, protested any destruction of the unique coastal forests. Even before Lees' death, though, a quiet revolution was happening inside that headquarters.

Robert Wilson became CEO of Rio Tinto in 1991: CEO of a company with an evil record of past treatment of both people and environment. Mines commonly mean decades of living with the same people in the same place, of breathing and polluting their air and their water, with eventual closure either in a landscape scarred beyond reclamation or just possibly in a landscape restored. Wilson was convinced that sustainable development was the new business model. Social, political and environmental contexts had changed since the bad old days of Rio's past.

Wilson convened a meeting of his peers, the CEOs of the other major mining companies. Many CEOs distrusted Wilson's initiative, but were gradually persuaded to work together for change. In 2001, they founded the International Council on Mining and Metals, with twenty-one of the world's biggest mining companies as members. The ICMM proclaims its promotion of sustainable development and accounting transparency.[5] This does not mean everything is perfect, but it is a very long way from the ruthless policies of the past. A related organization is the Business and Biodiversity Offsets Program (say Bee-Bop), uniting not only mines but businesses, government and NGOs like the Wildlife Conservation Society to compensate for local destruction with protection of similar areas elsewhere, aiming for net positive improvement for the ecosystem.

The titanium ore deposits of Fort Dauphin are among the richest in the world. They came from those mountains and that sea. The mountains were a part of Gondwanaland: granite and gneiss that erode tiny grains of heavy minerals. Then the sea, the endless pounding surf of the south-west trades and crashing storms from the south-east, centrifuged the dunes, and washed away lighter white quartz sand, concentrating the heavy black ore. The centrifugal spirals which the miners use to separate the sand are only the last stage in

5. ICMM, 'ICMM: International Council on Mining and Metals' (2013).

a long geological distillation of riches for the taking. No way could Rio hide, but no way would it walk away.

The upshot was that the QMM project in Fort Dauphin became Rio's flagship green mine: a new mine to embody the new principles of sustainability. In Fort Dauphin, Rio Tinto promised to do everything right.

The International Advisory Panel (the IAP) has given me a ringside seat. In 2000 Dan Lambert, president of QMM, invited three of us to become an independent panel on social and environmental aspects of the mine: Keith Bezanson, international development economist, Léon Rajaobelina and me, with the support of Jacques Gerin, who could not be a formal member because he had once been a consultant to QMM. By 1996 Léon had become head of Conservation International in Madagascar—a vantage point from which he could utilize his enormous network of government contacts without being implicated in the actual government. The IAP has met annually for ten years with reports open to all.[6]

At first I was extremely dubious about being involved with the mine. In 2002 Richard was invited to South Africa by Unicef. We took the chance to visit Richards Bay in Zululand, an ilmenite mine also owned by QMM. To our surprise, we were deeply impressed by the mine's commitment to the environment and to the welfare of its workers and their families. Nonetheless it would be a lot harder in Madagascar. The South African mine lies on seafront dunes with a forest adapted to regrow quickly after landslides, and thus after mining. The Madagascar site is on ancient fossil dunes where slow-growing hardwoods have adapted to the grudging soils. The origins of Richards Bay date to the apartheid era when

6. K. Bezanson, J. Gerin et al., 'Report of the International Advisory Panel on QMM' (2012).

Africans were meant to knuckle under and do what they were told, and then become grateful for post-apartheid improvements. Malagasy are at least supposed to be masters in their own land. However, the good intentions of the company seemed obvious. I signed up to the Madagascar IAP.

If I had changed my mind about the IAP, that was nothing to the change forced on Dan Lambert. IAP members stipulated that we could interview whomever we wanted: local and national government, NGOs, activists. Also that we would be free to say anything to anyone, individually or as a group. Dan agreed. He was so proud of all the good things planned by his company that he was sure we would be simply a propaganda coup to confront the environmental lobby. It was a rude shock when we came up with a whole raft of criticisms. Indeed, each of our copies of his proposed 2001 SEIA, the Social and Environmental Impact Assessment, were scribbled over, practically obliterating the text. Léon's was in technicolor, with highlights of red, yellow and blue for categories of error. Dan gulped, and admitted that it was rather a help to have something like informal evaluators turn up every year, whether or not he wanted to take our advice.

In particular we criticized communications with the local people, year after year, while QMM kept churning out ad-speak. Just as fundamentally, we kept advising the company to *limit* its benevolent interventions, with very clear messages about what it could and could not do. The temptation is to take over all the functions of local government because the company has the funding and organization to do so while government does not. However no company is rich enough (or benevolent enough) to run a whole region.

We didn't reckon, though, on the extreme underfunding of local government. For a while the *chef de région* was Harifidy Ramilson, the former head of the Upper Mandrare Basin Project. Dynamic and visionary, he pushed the region forward in spite of all the constraints. Then President Ravalomanana promoted him still further to be minister of agriculture. There was no funding, no backing for the political services: no alternative to help from QMM.

The company paid for a water plant for Fort Dauphin town. It now contributes to the entire regional water supply, though it is still managed by the national agency JIRAMA. QMM tried for years to find a way to coerce JIRAMA, also the electricity supplier, to pay its agreed share, but ended by supplying Fort Dauphin's electricity. It campaigns for drainage of puddles to conquer malaria. It supports the NGO that does family planning, HIV and STD testing. But, rich as they are, or were, they cannot do everything—and if people think they will do everything, they will be hated when something does not work. Fortunately, the World Bank and other organizations have decided to contribute to Fort Dauphin's hoped-for stability and prosperity. I will come to that later.

So far I haven't mentioned people who actually live around the forests and use its timber, charcoal, palm-leaf roof thatch, vines for lobster-traps, medicinal herbs. Romantics argue that the forests exist today because local people have preserved them since time immemorial. This is true. The forests are ancestral land. It is not very good agricultural land; it is no coincidence that the richest concentrations of ore lie under the areas still forested. Also the forests have always responded to rising and falling sea levels, shrinking or growing at their own rate, unrelated to human intervention.[7]

The situation has changed since ancestral times. Anyone who has been around for ten years has seen the hill-top rainforests retreat up to the precipitous slopes of the crests.

7. M. Virah-Swamy, 'Threshold response of Madagascar's littoral forest to sea-level rise' (2009).

The town of Fort Dauphin, some 50,000 people, has an insatiable demand for charcoal, the only general cooking fuel. The actual footprint of Mandena forest, the nearest coastal fragment to Fort Dauphin, did not change much from the 1950s to the 1990s, but its resources were increasingly thinned out. Missouri Botanical Garden has calculated that less than 10 percent of original littoral forest remains in only eleven distinct patches the length of the whole east coast, and that, at the rate it is being used up, it may soon be extinct as an ecosystem.[8]

Then there is poverty. Anosy and the other southern areas are among the poorest in this poor country in monetary terms. And regarding malnutrition, the area of Anosy north of Fort Dauphin has one of the highest levels of stunting and infant death in the country. The ways of the ancestors cannot support the present population.

This, of course, is an outsider's conclusion. QMM held many public meetings, with open discussion, to explain what benefits the mine would bring in spite of disruption to lives and landscape. People could not be expected to welcome the intrusion into their lives unless they were already of an entrepreneurial, not a traditional, mindset. And if I could not even imagine the physical and social dimensions of a titanium mine before seeing one in South Africa, how could rural villagers begin to do so?

All the calculations of whether the mine is good or bad depend on the counterfactual. If one thinks that people and forest will somehow muddle through before the hills are scraped as bare as Haiti, then there is no reason to think that money and organization will improve life. If one looks at the statistics of forest loss, one opts for the mine. And if one goes to a village tucked below the mountains, and climbs

---

8. Missouri Botanical Garden, 'Using aerial photography to measure Madagascar's littoral forests' (1995–2013).

to the newly cleared forest-edge fields, one may meet a man standing up, bending at a 30-degree angle from the hips, and planting a manioc cutting at his chest height in the newly cleared hill before him. He wipes off the sweat and admits he would much rather farm flatter land. He knows that the forest above holds the water for his rice-fields way below. But where else can he grow his manioc?

Clambering down the near-vertical hillside he cuts the sweetest, juiciest pineapple in all the world and gives it to the exhausted guest who has visited him in his new field.[9]

Oct. 30, 1989, Fort Dauphin.

The Villa Dally is inhabited by a gaggle of biologists preparing the environmental impact statement. The villa, an elegant structure with arched porches all round, is leased by QMM, a subsidiary of Rio Tinto. It overlooks a stretch of Fort Dauphin's main road from the airport that constantly reverts to white sand.

Ken Creighton came suddenly round the corner. He's a person one sees first by his smile. There are plenty of rangy, suntanned white men in white shorts in this country. This one smiles hello with every muscle in his face, with laughter lines from mouth and eyes that echo up toward his hairline like waves toward a beach. Ken hails from Russ Mittermeier's new outfit, Conservation International, and also the Smithsonian. He is heading what is undoubtedly the most intensive faunal and floral survey ever carried out in this country. The irony is that much of what they are surveying will disappear into the maw of the Canadian bulldozers. The next irony is that this could be the best thing that ever happened for the rest.

9. Ibid.

Ken unfolded marvels. Endemic rodents, like *Eilurus*, and three bats—the little *Myotis goudoti*; the horseshoe bat *Hipposideros commersonii*, with a squashed-up nose Ken compared to a mountain gorilla; the sucker-footed golden bat, *Pyzopoda aurita*, which walks up travellers' palm leaves using suckers on its wings like a flying tree frog. The last was known from half a dozen nineteenth-century specimens until Ken found it common at Ranomafana, and now, again, here. The bats, though, are a tray of dead specimens. The maps are almost alive.

I've never seen such maps. Once you get used to the idea that false-color forest is red, not green, you see a fairly dismal picture.

There are essentially three coastal forests on this corner of the island (not counting the rainforest on the mountain tops—I mean coastal forest on the dunes of white sand). They are all somewhat different, according to Ken and co., who have camped three weeks in each. Petriky is dryer, a transition between humid coast and dry western forests, crammed with plant species that exist no where else on earth. Mandena is nice but doomed. It is too close to Fort Dauphin. Every day trucks and people carry out wood. Word has gone out that it won't be protected, and so now it can't be. This was the forest where I first saw the US Steel Titanium research site with its ramshackle wooden scaffolding and the three-storey-high helical conveyor belt separating grey sand and white sand, back in 1975. The QIT-Fer Laboratory now is full of 2-foot-high mini test-spirals, busy washing white sand to the outside with grey sand left behind at the center, full of titanium.[10]

The third forest is Sainte Luce. Coincidentally it is beside another de Heaulme concession. It is 50 kilometers up the

10. A. Jolly, *A World Like Our Own: Man and Nature in Madagascar* (1980).

coast—and if I understand the implications of what they say, they'll recommend keeping Sainte Luce intact, and jettison the rest. Ken mumbled things like 'Only virgin stretch ... Thickest density of collared lemurs we've seen anywhere ... crested wood ibis... Our camp was here, between the lagoon and that drop-dead-beautiful bay... I kept waking up in the morning and thinking, I get *paid* to do this!'

Perhaps I should go see Sainte Luce.

Saturday, Nov. 4, 1989.

How do you describe an idyll? Turquoise sea foaming over granite rocks, palm trees and pandanus and picture-postcard beaches—it's all not good enough. Maybe a scientific description in micrograms of the endomorphin your brain would absorb to become as happy as it is existing among white beaches and turquoise sea. With no one else there but you and your friends and some reddish lemurs with grey heads, as though the world were created new this morning.

I climbed into the front seat of the Toyota Land Cruiser beside Monsieur de Heaulme. Madame de Heaulme courteously chose the back seat with the picnic, while Bénédicte and Claire de Heaulme and their young friends took a four-wheel-drive pick-up truck.

Monsieur steered us onto the execrable coastal road, and turned to me. 'What I want to talk to you about is how to save my portion of Sainte Luce as a nature reserve,' he said. 'You have done so much to publicize Berenty that now I hope you will take on the same task for Sainte Luce... My own property is very small: only 150 hectares, and 30 more hectares which were burned while I was in prison, when people thought they could do what they pleased. But my father and I have protected it for thirty-five years, which is why anything is left there at all.'

'Wait, Monsieur—I don't understand. Why did they burn it? To spite you?'

'No, no. For manioc fields. But it is white sand—even less fertile than the rainforest on laterite. It only takes one year for the very thin layer of humus to disappear, so then they move on. That's how the rest of the coastal strip has gone. I do not see QIT-Fer as the problem. They have been very polite. When I said there was no question of taking titanium from the my reserve area they did not even try to put in transects or take samples. They have quite enough deposits to mine elsewhere for many years. Alison, I am not against development. I am only against it on the Sainte Luce peninsula. Just to the north, beyond the village, there are hundreds of square kilometers of bare ground, the QMM concession, ready to mine.'

All this talk was between jolts as the Land Cruiser negotiated ruts, stones, bridges with holes, puddles 20 feet across. We stopped to buy bunches of ripe lychees and slowed to wave at grinning children in traveller's-palm-thatched villages. The road forked onto sand meshed by little tidal creeks with stone bridges consisting of only the central 10 feet of a 20-foot wash, so the Land Cruiser descended into the creek, crawled up onto the bridge bed, and back down into the creek like a terrapin hauling out to sun on a log. All around us stretched a sea of wiry grass dotted with traveller's palms.

We suddenly came to a beach with twenty big sea-going pirogues lined up beside that painfully blue sea, and granite islets fringed with white surf. The villagers were sorting the catch. One man walked off toward home holding a 4-foot shark by its tail.

The de Heaulme girls and I raced off to the shell of the first American Mission Boarding School at Manafiafy, the Malagasy and Lutheran name for this place. Three storeys, still solid, of enduring ironwood, with porches that face out through a coconut grove to the sea. It is a Malagasy school now with broken doors and no window glass. But some devoted teacher had written reading lessons on the board, and made a tidy square flower-garden of marigolds.

We ate sandwiches and lychees and drank Three Horses Beer beneath a huge shady badamier (a beach-almond tree, *Terminalia catappa*), and took a pirogue across a flowing lagoon to the de Heaulme peninsula.

Glossy leaves, pendant fruits, twisted short-stemmed trees that live in woods combed by constant wind. Six *Lemur fulvus collaris* to greet us as we entered: females red with grey heads, males bitter chocolate brown with luxuriant orange mutton-chop whiskers. Grunting and cocking their heads to watch us. We panted in time with their grunting as we forged up hill, along with whirring video and clicking cameras.

The best part of the wood for me was a grove of 20-foot-high pandanus trees. Madagascar has 100 kinds of endemic pandanus. Dark, cool, humid shade under the closed canopy of the forest, a stone's throw from the sea. Or maybe the best part was swimming in a pool sheltered from the surf by granite rocks...

There is said (by M. de Heaulme) to be buried pirate treasure. A *Treasure Island* film set? But why must everything be compared second-hand—to a book, a film, a picture postcard? I have walked through that forest today, and swum in the lagoon, and only just washed the salt off my skin.

Walking home at the waves' edge with the forested hill rising behind and the beachside pandanus clattering in the trade wind and the light golden on white beach and ever bluer sea, Bénédicte de Heaulme kicked the sand with her toes. 'There, Alison, do you see that dark grey streak in the beach? That's titanium.'

White sand and gray sand, the old English round:

> White sand and gray sand.
> Who'll buy my white sand?
> Who'll buy my gray sand?

The green flagship mine. QMM appointed Manon Vincelette, Québécoise forester extraordinary, to head their environment team. Her experience stretched from Madagascar to Quebec to the Congo. She set up nurseries of endemic and introduced trees. She drew up plans for restoration of native forest, also with quick-growing eucalyptus and casuarina for the charcoal burners. She hired the best and most passionate of Malagasy scientists. Johny Rabenantoandro of Missouri Botanical Garden headed up the botanical team, and Jean-Baptiste Ramanamanjato, herpetologist, the faunal side. Johny might have become an academic botanist and ecologist, but he was tempted by two thoughts. He wrote to me:

> Manon convinced me to join QMM in 2003, as responsible for botany, because of 2 strong opportunities to save littoral forest:
> 1. Permanent presence in the field (not like those projects that are often full of money but never exceed 5 years).
> 2. Ecological restoration (Madagascar will need this for the future generations and QMM was the only institution that was able to satisfy me on that).
>
> I also would like to stress the fact that Manon really trained me to be her Malagasy successor. It's an honorable behavior. Few expatriates were able to train a Malagasy successor. Often, they are replaced by a new expatriate. So now, as you said, I have become the head of the environment department.[11]

11. Johny Rabenantoandro, email to Alison Jolly, January 7, 2014.

Meanwhile QMM enrolled a social team of anthropologists and sociologists, most of them from Anosy itself.

I happened to be there the day, perhaps in 2000, that Manon confronted Dan Lambert, then the president of QMM, with the shock news that he could not mine all the ore body. She wanted 10 percent of it, the 10 percent with the very best forest, saved as local reserves. The scene: a clearing in Mandena forest, the nearest patch to Fort Dauphin, the first scheduled for mining. Dan, relaxed in short-sleeved shirt and field khakis. Manon, who for some reason is never bitten by mosquitoes, in her usual minimal halter top, short shorts, masses of waving curly hair, and steel-toed boots. White sand underfoot, a test spiral some 10 feet tall beside us, and eucalyptus planted only three years before overtopping the spiral, all under the glowing southern sun.

Manon said, 'You have to leave the heart of the forest. I have been planting the endemic trees for three years now. There are almost no fast-growing pioneer species. This forest cannot be restored from bare ground: there has to be 200 hectares of untouched woodland to build on.'

Dan said, 'You are talking about $250 million dollars worth of ore.' (This was at prices around fifteen years ago.)

'If you don't plan for a conservation zone, I shall see that you won't get your permit to mine.'

In fact, Dan was already half-convinced. It took more years to convince the local villagers, with conventions signed by them and by the local Water and Forests department about how the forest should be used. But in the dozen years since the use was limited to medicinal plants and traditional usage rights, I have watched the trees grow taller and the midstory and understory fill out. The conservation zone works. Similar zones are established as well for Petriky and Sainte Luce.

Rio Tinto has made two stunning policy gambles—so ambitious that I have serious doubts they can succeed. They have promised net positive improvement in both environment and society over the life of the mine. Much of the scrubbier forest outside the conservation zone is being razed. This means they pledge to offset the damage they cause by saving other forests... forests in use by people, with all the negotiations and trade-offs that implies. And they have pledged to never cause the extinction of a species. (That means vertebrates and plants. We still know too little about invertebrates.) To oversee this work, there is an annual meeting of a Biodiversity Committee of experts in the regional biology. A huge summary of Manon's team's own research was published in 2007: the academic resource of choice for Anosy littoral forest ecology. The team had discovered and named over twenty new species. Three of them are named after Manon: two plants and a scorpion. Is this last one a tribute?[12]

My favorite story is that of *Eligmocarpus*, a tree as determined as the California condors to become extinct. The last five specimens mature enough to fruit live in Petriky, where Andrew Lees died. The white, interlaced twisted stems rise to a crown 40 feet above you, holding dense black inner wood. If you take the tiniest blade of your Swiss Army knife to nick that white bark, it bleeds sap the scarlet of red blood. Little by little the sap dries to the crimson of coagulated blood. A tree out of the Snow White fairy tale: black and white and red.

The local name is *hazomainty*, black wood tree, which is also the name for ebony. A businessman asked for ebony, but when he arrived in Petriky he did not want *Eligmocarpus*. Unfortunately by that time local people had cut more than ten mature trees. Russ Mittermeier and I once roughly counted the rings of a fallen giant. It was some five centuries old.

Manon's workers patiently stripped the pods of surviving *Eligmocarpus*, finding perhaps one viable seed among a

---

12. J.U. Ganzhorn, S.M. Goodman et al., *Biodiversity, Ecology and Conservation of Littoral Ecosystems in Southeastern Madagascar, Tolagnaro (Fort Dauphin)* (2007).

hundred empty capsules. It took three years to find out how to make them sprout. The answer: ripe, freshly picked seeds, soaked for forty-eight hours before germination. In short, the tree can afford to wait through centuries until a cyclone arrives as the seed ripens: the only way you'd get forty-eight hours soaking on a substrate of thirsty white sand. Either that or it originated on better soil, further upstream. QMM was in more of a hurry. Now scores of little *Eligmocarpus* seedlings dot the conservation zone of Petriky. It is still far from out of danger: zebu eat the seedlings, and dissident villages may still hack down mature trees to spite the mine, which seems to value them so unaccountably.[13]

The World Bank and QMM played poker over financing the new port.

QMM's twenty-year exploration concession was due to run out in 2005. At that point it had to put up or shut up. It gained its environmental permit in 2001. The quantitative ore tests were done, the haul road course agreed, a mountain identified made of granite solid enough to build the outlet port—but QMM wanted a loan via the government to help build that port. That would implicate both government and the Bank in the project's success.

The original miners' plan was to build a port in a sensitive lagoon and estuary east of the town. A tube would carry ore slurry to waiting ships with no access for others. One advantage from the mine's point of view was that a peninsula conveniently shielded the estuary from the worst of the waves. Quite rightly, the environmental lobby was furious.

13. D.S. Devey, F. Forest et al., 'A snapshot of extinction in action: the decline and imminent demise of the endemic *Eligmocarpus* Capuron (Caesalpinoideae, Leguminosae(serves as an example of the fragility of Madagascan ecosystems' (2013).

The alternative was a port on the peninsula to the west of the town—a real port that could bring other business besides mining: exports, tourism, with vacant land behind to set up warehouses. Only it would need an 800-meter-long breakwater built to withstand cyclone waves.

The history of Fort Dauphin staggers from shipwreck to shipwreck. When French colonists first settled there in the 1650s, their great governor, Étienne de Flacourt, wrote that they already knew it was a terrible harbor. They judged it better to die of shipwreck beside Fort Dauphin's windy peninsula than of malaria at Sainte Luce.[14] In fact, that colony came to an end with the Massacre of the Pink Girls, when a ship bearing brides for Louis XIV's colonists foundered in the bay. The Frenchmen by this time already had local Tanosy wives, who took exception to the newcomers... one thing led to another.[15]

In 2005 when QMM was wooing the Bank, four different wrecks lay on the shore in various stages of disintegration, and one more in the middle of the bay. The storm swell from the south-east simply plucks a ship from its anchorage and parks it on Fort Dauphin's beach. A new port built to withstand the onslaught was clearly the only hope for future business expansion of Fort Dauphin, and the mine was the only hope for a new port.

QMM played coy, announcing that in spite of the millions of dollars already invested in mining surveys, their ilmenite was only marginally profitable. They could still walk away. The Bank played coy, pointing out that QMM was not releasing figures to calculate from, and that Bank and government had no reason to trust them. Finally in August–September 2005, both put their cards on the table. The project, the port

---

14. É. de Flacourt, *Histoire de la Grande Ile de Madagascar* (1661).

15. A. Jolly, *Lords and Lemurs: Mad Scientists, Kings with Spears, and the Survival of Diversity in Madagascar* (2004).

and a World Bank Growth Pole Project would all happen in Anosy.

Now, as of 2013, the Bank has resumed its Growth Pole Project, building roads. Fort Dauphin remains one of the most stable regions of Madagascar, so other agencies have decided to build on success: the EU with roads, Unicef with education. CARE was there all along, working with WFP to support food security. The small NGO Azafady, founded to counter the mine, now works in parallel. Civil society organization is also returning. At last there is some counterweight to the power and responsibilities of QMM.

Who benefited and who lost?

The three-year construction phase strained every inch of Fort Dauphin. An influx of skilled mining chiefs from South Africa, of mid-level workers from Southeast Asia, of road-building crews from Tana, and even employment for laborers from Fort Dauphin all caused prices for rooms and food to skyrocket. Of course anyone with a room to let or food to sell gained a bit of the boom. The very poor turned to aid agencies like CARE, or, as usual, to their extended families. After the boom, small businesses reaped profits as subcontractors, the areas right around the mine gained pig farms and honey hives, but there were groups of people who felt they'd been royally cheated.

There were villagers who had been moved to make way for the quarry. Some fishermen had to give up using one beach during the three construction years. Other fishermen's catches changed with the construction of a weir in the estuary. The numbers are minuscule: 63 households for the quarry village, some 70 fishermen by the weir, a few score more for the ones with the beach. QMM built new houses for the quarry village to replace the original tumbledown shacks. It introduced fiberglass outboard-motor boats with training for the fishermen, and set up schools, crafts and employment schemes. But the quarry people say the fields given them in compensation for their lost rice lands were no fair compensation; the fishermen say catch declined. Other farmers worried that they will be blocked from using the woods as they always have, and about general deterioration of soil and water.[16] And everyone wants employment. One of the chronic complaints is that managerial jobs go to 'foreigners,' not whites but plateau people from the capital with education.

The local branch of the National Office of the Environment (ONE) was supposed to sort out the specific claims, but since ONE nationally was broke, QMM has had to pay the independent government body with the resources to deliver an independent judgment.

For all the efforts of the company to describe its social policies as fair, transparent and helpful, there are critics who see no reason to believe such 'propaganda.' One fundamental problem, as always, is land. Land valued as the gift of the ancestors, inalienable, the only source of a family's rice? Or land which can be compensated by cash and new techniques?[17]

Dissidents from time to time block the road from mine to port. Whatever the rights and wrongs, people with grievances can clearly see that QMM would have the cash to pay them off—pay them over and over. Philippe Liétard, advisor to the government provided by the Bank, summed it up when he said, 'I have been in mining all my life. If you are a mining company, as far as the public is concerned you can't win.'

16. Andrew Lees Trust and PANOS, 'Madagascar: Voices of Change' (2009).
17. Rio Tinto, 'Rio Tinto QIT Madagascar Minerals' (2013); C. Seagle, 'Inverting the impacts: mining, conservation and sustainability claimes near the Rio Tinto/QMM ilmenite mine in Southeast Madagascar' (2012).

September 13, 2012.
Sainte Luce is now scheduled for hurry-up mining. Not the tiny de Heaulme reserve, but the sprawling area long slated for mining with its own conservation zones and offsets.

Léon is in Singapore, being made chief of Conservation International's Africa region. Keith and Jacques fight shy of the 2½ hour journey on a horrible road. They opt for interviewing Lisa Bass, the new head of Azafady. The Azafady NGO was originally founded to fight the mine—now they have settled instead to working in parallel with QMM within the Sainte Luce community.

Richard and I want to see it all for real! An ebullient Azafady staff member with the improbable name of Forrest Hogg accompanies us—and Monsieur de Heaulme sends a guardian along to unlock the boats for the de Heaulme peninsula of glorious beach-front forest. I am *not* going to Sainte Luce without a chance to swim off that beach.

It is an ever more horrible road. The 4×4 pick-up truck does embarrassing things to my bladder. But the little roadside villages are still postcard-pretty: stick walls, floor and tiny front porches on foot-high stilts against rain and rats, palm thatch down over their eyes, shielding the front porch. Papaya trees and pineapple plants. Grinning, waving children everywhere.

Halfway there we pass a pick-up heading the other way. 'Stop!' I demand. 'Stop! That car is labeled Missouri Botanical Garden!' The driver of the other car hops out. Reza Ludovic, community facilitator and botanist. Reza is surveying people's use of beach-front forest. This isn't the mine forest, but an 'offset,' a forest off the ore body which QMM is pledged to conserve in lieu of its destruction of other forest fragments.

Beach-front forest grows on white sand, or in this case white sand with streaks of black, titanium-rich sand grains.

Feathery fronds of palms, tough leaves of pandanus on candelabra stems, as well as the noble crowns of the badamier and the powder-puff tree, both of which give shade on tropical beaches around the world. Being Madagascar, most of the palms and all of the pandanus are endemic species, as are many of the other trees caressed or hammered by the wind from the sea.[18]

I privately call beach-front forest 'The place with no loo paper.' Without exception its leaves are tough, glossy, drought and salt-resistant. There isn't even one undergrowth plant which has soft tender leaves fit for the purpose.

Reza is particularly worried about the fanola mena and the fanola fotsy, otherwise known as *Asteropeia micraster* and *Asteropeia multiflora*. They belong to an endemic genus. Some (but not all) botanists credit that one genus with the distinction of being a whole separate plant family—endemic, of course. Nearby communities use only native forest trees for firewood. These two are known as the petrol trees: so full of oily resin that they catch fire instantly. *Asteropeia micraster* is already on the IUCN red list of endangered species.

House building: the first post in the forward porch must be fanolamena; then other trees from the beach forest like uapaca or hintsy make the rest of the supporting uprights. Uapaca species are tall with stilt roots. Hintsy is the orange teak-wood tree I watched being logged in the Masoala back in 1985. They are slow growing, like all hardwoods, so it doesn't seem worthwhile to most folk to try to plant more. The forest has always provided.

When the house is finished and thatched with palm leaves, Madagascar periwinkle provides a blessing. Thank goodness the periwinkle is a pretty weed, a Madagascar

---

18. Badamier, beach almond: *Terminalia catalpa*. Powder-puff tree: *Barringtonia asiatica*.

endemic that now grows worldwide and isn't threatened even in its home country in spite of its harvesting as the Western cure for childhood leukemia.

But after talking to Reza, I am even more worried. The 'offsets' are supposed to be community-managed forests with true sustainability under a Rio Tinto umbrella. No way can Rio Tinto afford alternative social support which would save even this stretch of woods, the glorious beach forest that stretches from Sainte Luce almost down to Fort Dauphin. The biggest of the offsets, on the whole mountain range behind the coastal plain, is a much bigger problem. Reza says the beach communities all own land up on the hills where they can grow more than meager cassava, the only domestic crop which extracts a living from white sand. (I did see a pineapple field on sand, but with the pineapples spaced a meter apart, not fat and juicy and growing higgledy piggledy the way they grow on the hills.) The *tavy* that is clearing the hills almost to their summits comes not just from hill villages, but from all the villages below. Not to mention illegal rosewood logging.

At least QMM is aware of all this. Reza reports back to Johny Rabenantoandro as head of QMM environment. Reza, an impressive manager as well as a botanist, has long experience in reconciling communities and forests.

I am obviously too creaky to get down on the ground on the mat the village spreads for us to talk with Reza, at least not without the indignity of assembled strong men hauling me back onto my feet. A very old lady brings her very own small wooden bench from her house for me to sit upon. Lovely gesture. Lovely people. At least most of the time!

'Vola!' harangues the woman in red at the sweet-potato stall. 'Vola! Money! What we want and need is money!' I crossed the lagoon bridge on foot. The bridge is now forbidden to QMM

cars on grounds of safety. Silly. Any driver worth his salt here can keep his wheels on two longitudinal narrow planks. It is irrelevant that all the cross-planks in the middle have caved in. In the lagoon a small boy sits astride a plank, paddling it around the still brown waters with an oar. Above him the coconut palms are tall enough to catch the wind from the sea. The palms shake their fronds like petulant teenagers.

Richard and Forrest walk on toward the sea-side village. I opt to sit at the sweet potato stall by the bridge, with Paola of Azafady as interpreter.

'Money!' translates Paola, as the stall-holder in red offers me a present of two boiled chunks of sweet potato. 'The woman says "QMM is already taking soundings, taking samples of our land. We don't know why, we don't know where the samples go. So QMM should already be paying us money for the pieces of our land!"' Paola isn't really rising to the oratory. The voice of the woman in red rises and falls, with emphasis in wiggling shoulders and hips. At one point she says no one in the area is trained for a new life; all they know is fishing and hewing wood. I get the gestured pulling in of nets, the chopping wood at ankle height.

'But we do know how to spend our money!' she claims. 'QMM doesn't need to give us different compensation. If we had money we'd know ourselves what to do with it.'

I ask, 'Supposing you did have money. What would you do?'

She looks into the middle distance, thinking. Her loud oratorical voice drops to the level of a dream imagined. 'First, I would buy land. I would buy some land. And then I would build a house on my land. A real house, made of bricks, not like all these houses here… And then I would improve my stall into a real store.'

QMM has had endless trouble with people compensated in cash for loss of land or fishing rights. People with no experience of such a windfall tend to spend it on rusty taxis that break down, or simply on the flock of distant relatives who suddenly descend on them. Or booze. And then they come back for more money, saying QMM has cheated them of their land. I read an article somewhere about lottery winners in England and the USA, and how the big win can lead straight into bankruptcy. Not so very different.

I say: 'QMM will do much, but I don't think they will give money.'

A gray day at the beach.

I have so many memories of picnics here with the de Heaulme family, arriving in shiny Land Cruisers with an outboard motor boat waiting to take us across the lagoon mouth, then unpacking the hampers of saffron rice salad and grilled lobsters...

Not so now. Today we've rounded up two pirogues and their oarsmen. I am put on a narrow cross-thwart, Richard on another. The water is an inch below what would be the gunnels if it were a boat worthy of such vocabulary ... in other words below the hacked edge of the tree trunk we are precariously sitting on.

This doesn't seem to bother our paddlers.

We start quite far up-lagoon, which gives us a great view of women and children fishing for crabs among the mangroves with the bed-nets that are handed out to prevent malaria. Daily protein matters more than the chance of illness. Everyone waves happily.

We land on the peninsula. And walk through to look out to the castle rocks that shelter the swimming beach. Surf crashes mightily on the castle rocks. Our son Arthur once took a glorious picture of a man standing on those rocks, with blue sea and paler blue lagoon and surf-splash beyond twice as high as his head. The man was no picnicking tourist. He was fishing mussels at low tide to use for lobster bait.

But it is gray today, and empty. Richard and I walk round the curve of the beach to where a fallen tree trunk gives hand-holds for getting in and out of the water. The dreadful thing about age is that it takes longer to get undressed than the actual swim. You have to hang on to a branch while pulling your trousers off. We manage. I paddle about with Richard in the warm sea. And then I shake the sand from dropped clothing: black sand and white sand, quartz and titanium.

Is this the last trip to Sainte Luce, the last view of the perfect beach and waxen forest? On this trip to Madagascar I am saying goodbye forever. To my lemurs, to Berenty and Fort Dauphin and Hanta and the de Heaulmes. And the forests…

June 2013: Sainte Luce mining is cancelled for now. The prices of titanium and other minerals have crashed on the world market. Rio overextended itself about three years ago with the purchase of the aluminum company Alcan. Having hurried up plans for mining in Sainte Luce, it is all on hold. They have enough to do with the mining already started at Mandena. The woman in red by the sweet-potato stall can whistle for her money.

But they swear that the environmental and social programs already started in the region will not be cancelled. The president of QMM, Ny Fanja Rakotomalala, has appointed a new head of social actions: Lisa Gaylord. Lisa was the chief of the USAID biodiversity program from the first beginning of the Environmental Action Plan in 1991 until USAID pulled out after the 2009 coup. I wrote in one of my other diaries:

> I think Madagascar should give Lisa Gaylord its medal of honor for twenty years of the highest service to this country. I'd say a statue, but never, never, never could one raise a statue to a quick blonde American, even with a Malagasy husband and kids. Maybe more appropriate, name a forest saved, or a waterfall, or the next new species of quick blond lemur...

Lisa chose to head the social team at QMM rather than leave the country of her family, so she is now applying her accumulated expertise to the landscape of Anosy.

Nov. 28, 2013, Thanksgiving Day.
The QMM Biodiversity Committee meeting in Paris. A standard hotel conference room: a square of tables with the three-person QMM team on one side, and five scientists on the other, Jon Eckstrom paid by QMM.

Ny Fanja Rakotomalala is now president of QMM. A smooth-faced and smiling man. You wouldn't guess that on his shoulders as a mining engineer, he is in charge of turning profit from a new mine with all its teething problems, and now a drop in worldwide mineral price. Beside him sits Johny Rabenantoandro, thin, intense, and in quiet moments sentimental about his people and his forests. In many ways he would rather be following his training as Missouri Botanical Garden's brightest coastal forest specialist instead of working with a mine—but he has become a formidable administrator, Manon's successor on the environmental side. The amount of biological work his team accomplish is staggering. Johny Rabenantoandro now has a further mandate: as of 2013 to organize all the water supply not just for the mine, its dredging lake and settling ponds, but even the drinking water of Fort Dauphin.

And then Lisa Gaylord, blond curls bouncing. She grapples with the ever-unsolvable problems of the populace. She has moved to a landscape approach here, as with USAID: one absolutely has to look at the whole regional picture and base it on land management. She admits, though, that

> Even with all the commitments from Rio Tinto towards the Net Positive Impact on Biodiversity in Anosy, if we do not have the commitment of the local government and communities, I can guarantee to all environmentalists (including myself) that we will not succeed in the long term. This is where I will continue to put my efforts to create a critical mass of people in Anosy who understand and can help promote this mutual engagement as we did over the 20 year Environment Action Plan that continues to ensure that our collective results remain sustainable. It is an incredible challenge, but we each need to move forward to respond to the complexity of the challenges.[19]

Ny Fanja cheers me as he outlines the increasing cohesion and power of the mines. He is chair of the Madagascar Chamber of Mines, MCM, comprising eighteen large and medium-sized companies. In September 2013, the MCM hosted a conference of MIASA, the Mining Industry Associations of Southern Africa. So far thirty different mining companies have signed up to EITI, and thus to transparency in their financial dealings with government. Six private-sector groups, including MCM, are working with the World Bank to draw up a private sector charter. Though Ny Fanja phrases it more diplomatically, the companies are fed up dealing with an unstable and kleptocratic government and are slowly uniting to launch their own agenda.

I briefly think 'Déjà vu all over again.' I remember so well when, as architects of the Environmental Action Plan, the Bank and the donors put together a charter of what we

19. The phrasing is from an email from Lisa Gaylord, January 2, 2014. I've put it into the diary to use her own words. Lisa collaborated on a major policy document stating USAID's approach: USAID, *Nature, Wealth and Power: Leveraging Natural and Social Capital for Resilient Development* (2013).

thought Madagascar needed. The difference, though, is that elite Malagasy were mystified by our fascination with little lemurs. This time around the elite know all too well that they need companies to invest.

Meanwhile, as I have always warned, a new CEO of Rio Tinto is slashing personnel and retreating from overall company commitment to biodiversity. I've feared this for years. Environmental groups take note! Ny Fanja ensures us, though, that QMM is safe—still the flagship, still the most sensitive of Rio's mines, though far from the largest. It has many challenges. Mineral price volatility: titanium ore is way down right now. The independent monitoring agency, ONE, the National Office of the Environment, is still starved of funds so it is difficult to maintain credible oversight. And when local people complain, there is the complicating factor that the mine is the only body worth complaining to—the only one that listens or has the resources to respond. But all the environmental and social programs remain in place.

The Paris meeting is cost-saving. Cheaper to fly Ny Fanja and Johny and Lisa this way than six of us in the other direction. That meant I could come as well. No way could I make it to Madagascar any more. Paris is my last trip. I am glad to end off with a flourish of my double canes at Eurostar.

I get to sound off all my opinions. Also to say goodbye to friends—incidentally buttonholing dear wise Jörg Ganzhorn to tell me about Tsimanampetsotsa Reserve for my book! Alsatian beer with Jörg still has its charms. (He favors low-alcohol 'white beer.' I can't see the point.) Maybe, though, I'll have no appetite ever again. I couldn't go wallow in Pete Lowry's very favorite Parisian restaurant. But, in return, we all recommended that Missouri Botanical Garden, Pete's Lowry's outfit, start administering the Sainte Luce area in collaboration with its communities. Reza Ludovic will head the project.

Those forests are more and more amazing! If anyone ever asks why the struggle to conserve is worth it, Pete says the forest of Bemangidy, one of the offset sites, a remaining tongue of low-altitude rainforest, has yielded twelve new plant species this year. They are, of course, critically endangered already because almost all the rest of the hills in Anosy are scraped back well above the level of Bemangidy. Thirty more new species wait in the wings for taxonomists to describe them. One of the latest, a tree hung with clusters of elegantly pendant pink and white flowers, is *Bemangidia lowryi*.[20]

Conclusion.

I have always said that the QMM presence in Fort Dauphin is a gamble. It does seem to me the only possible way to acceptable development. Anosy needs the deep-water port, the infusion of money into local businesses and villages, and the goodwill of a company pledged to do its best. Still, it could simply become social upheaval and crashing disappointment. To make it work, the company must continue to honor its pledges, local government must actually govern, and local people must see into the future—Ny Fanja stresses that without others QMM can never do it alone. Also environmentalists must continue to champion the environment. The moment the outside world takes its eyes off Anosy a future CEO who is not a choirboy could bury the local people while he digs up the titanium.

As the national government staggers on toward the 2013 election, the mining sector has moved into the place that the conservationists once claimed, shaping government policy. Which may not be all bad. A gamble for the highest stakes of all: the future of Madagascar.

---

20. *Bemangidia lowri:* L. Gautier, Y. Naciri et al. 'A new species, genus and tribe of Saptaceae, endemic to Madagascar' (2013).

TWENTY-FIVE

Where are we now?

January 14, 2014.

I will never go back to Madagascar. I'm home in England surrounded by loving and worried family. Somehow dying doesn't worry me at all, at least not yet. I've had lots of fun and excitement in life. Richard and I have just celebrated our fiftieth wedding anniversary! I told our 12-year-old granddaughter that now I feel I'm lying in a small boat that drifts with the current down a calm stream, maybe with willows overhead. I might sit up and find an oar and try to paddle the other way, but why bother? Only I still wish I knew what happens next...

Russell Mittermeier is going strong, ricocheting round the world for his brainchild Conservation International. This year he and his collaborators brought out a comprehensive strategy for what would be needed to save all lemur species. At least we know now what ought to happen.[1]

This past Monday Russ had dinner with the equally indefatigable Patricia Wright at the Carleton (ex-Hilton) in Antananarivo. Pat has since zoomed back to Ranomafana with yet another television team. You will soon see her and

1. C. Schwitzer, C.G. Mittermeier et al. *Lemurs of Madagascar: A Strategy for Their Conservation 2013–2016* (2013).

my dear friend Hanta Rasamimanana standing three storeys tall on an IMAX film, *Madagascar,* opening near you in April. At least, if you live in the developed world. Madagascar itself does not have IMAX. Ranomafana's research goes from strength to strength. In August it held the last of the congresses that frame this book: an International Prosimian Congress. The big change is that most papers are by Malagasy speaking on their own biodiversity, eager to advance their own careers in conservation. A contrast to the continuing bewilderment of so many other Malagasy as to why anyone would want to visit forests! And a huge swing from all the meetings in the past dominated by foreigners.[2]

Hanta, who was a program chair for the August conference, is gearing up for yet another major research blitz on Berenty this coming April, with three other long-term researchers and half a dozen students, Malagasy and Canadian. This time it is vegetation and a management plan to maintain the forest, requested by Claire de Heaulme Foulon. Claire and her husband Didier have taken on the family responsibility of conserving the Berenty Reserves for all their different peoples and trees and lemurs, while Claire's sister Bénédicte runs the sisal plantation. Hanta also carries on the Ako project of storybooks for Malagasy children on top of everything else she does!

Frans Lanting is recognized as one of the world's great photographers. His vision is so unique that when a page of the *National Geographic* falls open you recognize his work in an instant. His *Life: A Journey though Time* is a thrilling performance, where Frans's photos trace the rise of biology from the first bare rocks to the exuberance of the present day, while a whole live symphony orchestra plays music composed by Philip Glass.

2. Valbio, 'ICTE–Centre Valbio publications' (2013).

Léon Rajaobelina continues as director of Conservation International's Madagascar branch in Antananarivo. This has been his base as an *éminence grise* in politics, and in mining policy, and dozens of other activities. Since he knows absolutely everyone who counts in Madagascar, I don't expect he'll ever stop.

Joe Peters left development practice in anguish after the debacle at Ranomafana. He became an academic, then a performing songwriter. His farewell concert was last Friday along with his and Dai's thirty-fifth wedding anniversary. Then on Monday Dai left for an early morning flight to her latest agricultural consultancies, this time in Haiti and Nigeria. Dennis del Castillo, agronomist, has worked for the World Bank, USAID, and is now with the Peruvian Amazon Research Institute.[3]

Alison Richard has retired as vice chancellor of Cambridge University and is now back at Yale. She and Joelisoa Ratsirarson have emerged from Bezà Mahafaly as I write. The summary:

> Just back in Tana. LOTS of rain, sifaka, maki, greenness, wasps, mud, lots MORE rain, scorpions, sira tany [salt of the earth], the forest bursting with life, fields full of maize and manioc, wading across the Sakamena waist high, even more rain STILL, hot nights in my tent, much accomplished, everyone of good cheer.[4]

And this is good cheer all across the south. Good rain in January means food for all this year.

Most of the other people I've named are retired, though they go right on being involved with Madagascar, whether Malagasy or other. Jean-Jacques Petter died in 2002 full of

---

3. Joe Peters, email to Alison Jolly, January 8, 2014; www.reverbnation.com/joepeters, http://amazonecology.wordpress.com/2011/05/06/connections-profile-dennis-del-castillo.

4. Alison Richard, email to Alison Jolly, January 24, 2014.

honors, survived by Arlette whose own studies of lemur reproduction has been the foundation of so much later work. Wonderful Madame Berthe has also died, mourned and honored by the whole GERP, the Groupe d'étude et de recherche sur les primates de Madagascar, the Malagasy primatological society she founded, which hosted the International Congress of 1998.

As for what happens next, all of Madagascar is waiting for the presidential election results. No, to tell the truth, it is January so most Malagasy are into the serious business of planting rice and manioc in the rains. Meanwhile the politicians enter their end game…

For the last five years Madagascar has waited for elections. When the final round of votes are counted, promised for January 22, 2014, the country will have an internationally recognized government. At least, if the loser accepts defeat and does not choose to plunge the country again into chaos.

The Transitional (post coup) Government had no incentive to agree to elections while their fingers were in the till. They used the ancient Malagasy ploy of saying 'Yes, yes, yes.' Then they fixed meetings with the opposition and cancelled them at the last moment. Or else they imposed conditions which made others cancel, such as: 'If you set foot in Madagascar we arrest you.' Foreign pressure toward a real government grew more and more intense. A committee from the Southern Africa Development Community tried to bring the government to heel with increasing frustration. However, the pressure continued. When foreign aid was abruptly cancelled after the coup in 2009, it amounted to a loss of 40 percent of the government budget, which gives a good deal of leverage! Now at last there is the promise of stability.

Magistrate Béatrice Atallah took on the responsibility of organizing elections. If it weren't so serious, it could play as farce. First round April. No, July. No, October.

But then, I am American. I have no right to be sarcastic coming from a country which amply demonstrated in 2013 how to play politics as farce.

Much has changed already. A recognized government means it is worth signing agreements that will be honored. The World Bank and other donors have re-engaged. The Chamber of Mines, described in the last chapter, which had long existed, is now reactivated.

All agreed that former presidents Marc Ravalomanana and Andry Rajoelina could not be candidates. But that did not exclude Ravalomanana's wife or Didier Ratsiraka. Rajoelina, leader of the coup and president of the Transitional Government, announced that if those two stood he would put his own name down as well. Amazingly, Magistrate Atallah and some tough negotiators cleared them all away. She is a lady of presence. (I met her only once in an airport: black dress, rakish black hat and masses of gold jewelry. Presence felt in all directions by those who backed away in respect and those who rushed up to claim acquaintance.)

The first round in October posted forty-one candidates, all orating on their own radio stations to a populace that does not read posted campaign papers—because most don't read. Still, international electoral observers conclude the election itself was carried out fairly.

This led to a run-off on December 20th between the two leading candidates: Jean-Louis Robinson and Hery Rajaonarimampianina. Robinson was the candidate of Ravalomanana's party, pledged to continue his policies of an open economy, environmental support, and health and education. Rajaonarimamapianina (call him Hery) carries on the Transitional Government. However, since his party has caved in enough to hold elections the economic policies are predictably also going to appease investors and donors.

On January 7th, 2014, the Electoral Council announced Hery the winner, by 53 percent versus 47 percent. A series of questions about who financed each candidate has ricocheted round the Malagasy press. No one has actually fingered rosewood profits, but in every previous election rosewood exports rose, though never to the levels of 2009 and 2013. Outside observers may declare an actual election fair, but they cannot trace the tortuous paths that led to the final rolls.[5]

Naturally Robinson wants a recount, alleging fraud by the ruling party, but he has announced he will abide by the Council's final decision if Hery's side will promise the same. We do know that when the dust settles, Madagascar will have a new recognized government but be facing the same old problems.

January 20, 2014.

On Friday, January 18th, the Electoral Commission confirmed Hery Rajaonarimampianina as president. Questions remain: where did he get $43 million in campaign funds? Was it legal for him to block his opponents' Land Rovers and T-shirts at the port, let alone past presidents' entry into the country? Was it indeed legal for Andry, excluded from presidential campaigning, to appear on the campaign posters for Hery and his party? Now it seems to be legal.[6]

Well, let him rule legally and well. This is his chance.

Not all is good. There is widespread breakdown of law and order, or feared breakdown. This is wildly exaggerated by

5. Anglo Malagasy Society *Newsletter*, December 2013.

6. M. Andriamananjara, 'Madagascar finally has a new president, but uncertainty remains' (2014).

the foreign press if it involves a visitor. The stoning to death of a tourist on the vacation island of Nosy Be sent shudders. Villagers accused him of causing the drowning of a local boy. I suspect this is an isolated case of fear turning to paranoia. That has happened sporadically before. It is exactly the kind of reaction that Eleanor Sterling feared so long ago on the idyllic beach of the aye-ayes. But violent robberies have clearly increased. The underpaid and undisciplined police are more likely to arrest people rich enough to bribe their way out of jail than to go after killers.

Zebu rustling in the south has robbed the people of their bovine bank accounts. I haven't been able to find out where the cattle go. Traditional roots are deep: a century ago, one of a man's wishes to his newborn son was 'May you go strong and steal many cattle.' Now, though, it is on an industrial scale: someone is buying up thousands of cattle on the hoof. The increase in gun crime has led some embassies to recommend no one takes the Route Nationale road to Fort Dauphin!

In the south as well the 2013 rains were topped off by a cyclone—ideal for breeding migratory locusts. Again it is foreign donors who find funds for anti-locust campaigns. Famine does not arrive on the black horse of the Apocalypse. It descends from the sky on glittering wings.

Rosewood cutting and smuggling have soared to the levels of 2009. Guy Suzon Ramangason, director of Madagascar National Parks, this week stated publically that the final destination is China, while the government fails to intervene. 'There is a network of *Mafiosi* of bois de rose. Money in this type of network is very very powerful.'[7] Do you remember his namesake Guy Razafindralambo declaring 'If we destroy Madagascar and turn it into a desert, what will we do? We'd have to swim!'

7. T. Ford, 'Madagascar's trees vanish to feed rosewood trade' (2013), p. 1.

But many Malagasy keep working for the good of their country in the civil service and other positions. The re-invigorated Chamber of Mines is one kind of hope. Besides the stories I have already told, people and organizations who have been working in the country for decades are still there: WWF, Conservation International, Wildlife Conservation Society, Durrell Wildlife Conservation Foundation, Missouri Botanical Garden. Many newer ones have appeared: Mitsinjo, which plants trees in Andasibe where the indri sing; Fanamby, founded by Serge Rajaobelina (son of the inspirational Léon); Blue Ventures, which works with octopus fishermen to set up no-take zones; Madagasikara Voakazy, which started as Bat Conservation; the Japanese Croix du Sud, which works with villagers to replant spiny forest trees cut for timber; and a host of others. The GERP itself, the primatological society, undertakes consultancies.

The biggest mind-change though: in September the Wildlife Conservation Society and the government of Madagascar announced that 705,588 carbon credits are certified for sale from carbon stored in the 375,000 hectare Makira Forest. That is 32 million tons of CO_2 potentially saved as rainforest, if firms looking to offset their carbon emissions actually buy the carbon credits! The Makira is a huge tract linking the Zahamena Park of rainforest on the eastern mountain chain with the Masoala Park, home of the red ruffed lemurs where this book began in young Martial's slash-and-burn clearing. Some 240,000 people live or depend on the Makira. It continues to be their home. It is called a *Natural Park*, not a *Reserve*. The project has long been in the making: mooted in 2001, agreed in 2008 just before the coup, and now at last triumphantly placed on the carbon market.[8]

8. Wildife Conservation Society, 'Madagascar puts first-ever government-backed carbon credits on open market' (2013).

Furthermore, Conservation International is looking for carbon credits for two more great swathes of eastern rainforest: from Sahamena southwards to Andasibe, and the Ranomafana National Park corridor to the mountain massif of Andringitra.

REDD, 'Reducing Emissions from Forest Destruction and Degradation,' is controversial. For many, this scheme, where companies buy credits and pay for forest saved, is the best hope to actually recompense owners of forests for conserving a resource that is precious to the whole planet. The catch is the word 'owners.' If an organization wishes to buy carbon credits, it needs assurance that the sellers have secure, verifiable tenure. That means in turn that only large companies or umbrella organizations are in a position to sell. No way could 240,000 villagers cobble together the right to pledge conservation of their own land. The government is backing the carbon credits, but WCS and CI have to negotiate the deals.

In fact this raises every question we have faced in the course of this book. WCS has spent decades working with the Makira communities as well as calculating forest carbon stocks and ongoing forest degradation. Will this huge project resemble Bezà Mahafaly, where years and years of personal contact have generated real trust? Or do the villagers accept WCS as they do QMM on a whole range from resentment to collaboration, because the outsiders are clearly not going away? Even worse, I am always dubious of huge top-down schemes in Madagascar. Too often they never reach the people at the bottom they were designed to help! Worse, might an influx of quick cash to both villages and government generate the local greed of a dysfunctional family squabbling over a will?

As you will have gathered, I am convinced that development is necessary. The counterfactual of no development is a downward spiral of degradation for environment and society. However, it all depends how development is done.

❦

At the beginning of this book I wrote of the continuing battle between three views of nature: the aesthetic/scientific; the traditional; the economic. Economic arguments win, at least so far—perhaps with REDD to support them. Nonetheless the scientific and aesthetic idealism of foreigners and Malagasy alike lies behind all the efforts. Love of nature is a deep river, flowing under its turbulent surface.

I admit that Madagascar isn't *important*. It isn't important to the whole earth like the vast swathes of the Amazon or Indonesia with their carbon stocks, their millions of interlocking species, and their rampaging palm oil plantations. It isn't important even symbolically like the greatest living land animal, the elephant, butchered for Asian trinkets.

It *is* important, though, as its own alternate world of evolution: how to build a rainforest or a spiny forest or a baobab forest. It is the ecological theater and the evolutionary play with an entirely different cast of characters.[9] It is important as the only home of its forest species, birds and trees and insects and frogs and the hundred-odd kinds of lemurs that live only here, from the red ruffs of the rainforest to white sifaka leaping among the spines, from Madame Berthe's tiny mouse lemur to the indri who sing from hill to hill. And, of course, my very own swaggering, pugnacious and ever so maternally doting ring-tailed lemurs.

Madagascar is important above all as a test case for any ideal of sustainable peace between humanity and nature. If humanity, Malagasy and outsiders together cannot save Madagascar, what hope is there to save the planet?

---

9. G.E. Hutchinson, *The Ecological Theater and the Evolutionary Play* (1965).

References

Albignac, R., G.S. Ramangason et al. (1992). Éco-développement des communautés ruralles pour la conservation de la biodiversité. Paris: UNESCO.

Alexandre, C. (2007). *Violences Malgaches*. Antananarivo: Foie et Justice.

Andrew Lees Trust and PANOS. (2009). 'Madagascar: voices of change.' http://andrewleestrust.org/Reports/Voices of Change.pdf.

Andriamananjara, M. (2014). 'Madagascar finally has a new president, but uncertainty remains.' *Global Voices*.

Andriamananjara, S., and A. Sy. (2013). 'It's time to break the "Madagascar Cycle".' *Africa in Focus*. Washington DC: Brookings Institution.

Anglo-Malagasy Society (2013). *Newsletter* 82, December: 1–8. London: AMS.

Bezanson, K., J. Gerin et al. (2012). 'Report of the International Advisory Panel on QMM.' www.riotintomadagascar.com/english/summary.asp.

Bishop, M. (1954). *A Bowl of Bishop: Museum Thoughts, and Other Verses*. New York: Dial Press.

Bishop, M.G. (1962). *A History of Cornell*. Ithaca NY: Cornell University Press.

Boserup, E. (1965). *The Conditions of Agricultural Growth*. Chicago: Aldine.

Brown, M. (1995). *A History of Madagascar*. London: Damien Tunnicliffe.

Burney, D.A. (1996). 'Climate change and fire ecology as factors in the Quaternary biogeography of Madagascar.' In W. Lourenco (ed.), *Biogeographie de Madagascar*. Paris: ORSTOM, pp. 49–58.

Burney, D.A. (1997). 'Theories and facts regarding Holocene environmental change before and after human colonization.' In S.M. Goodman and B.D. Patterson (eds), *Natural Change and Human Impact in Madagascar*. Washington DC: Smithsonian Institution Press, pp. 75–92.

Burney, D.A. (2003). 'Madagascar's prehistoric ecosystems.' In S.M. Goodman and J.P. Benstead (eds), *The Natural History of Madagascar.* Chicago: University of Chicago Press, pp. 47–51.

Cai, W., S. Borface et al. (2014). 'Increasing frequency of extreme El Niño events due to greenhouse warming.' *Nature Climate Change* 4(2).

Conable, C.W. (1977). *Women at Cornell: The Myth of Equal Education.* Ithaca NY: Cornell University Press.

Cornia, G.A., R. Jolly et al. (1987). *Adjustment with a Human Face.* Oxford: Clarendon Press.

Corson, C. (2011). 'Territorialization, enclosure and non-state influence in struggles over Madagascar's forests.' *J. of Peasant Studies* 38(4).

Corson, C. (2012). 'From rhetoric to practice: how high-profile politics impeded community consultation in Madagascar's new protected areas.' *Society and Natural Resources* 25: 335–51.

Dammhahn, M., and L. Almeling. (2012). 'Is risk taking during foraging a personality trait? A field test for cross-context consistency in boldness.' *Animal Behav.* 84: 1131–9.

Decary, R. (1969). *Souvenirs et croquis de la terre malgache.* Paris: Éditions Maritimes et d'Outre-Mer.

Darnell, E., and T. McGrath. (2005). *Madagascar.* Glendale CA: DreamWorks Animation.

Devey, D.S., F. Forest et al. (2013). 'A snapshot of extinction in action: the decline and imminent demise of the endemic *Eligmocarpus* Capuron (Caesalpinoideae, Leguminosae).' *South African Journal of Botany.*

Dew, J., and P.C. Wright. (1998). 'Frugivory and seed dispersal by four species of primates in Madagascar's eastern rainforest.' *Biotropica* 30: 425–37.

Dewar, R.E., and A.F. Richard. (2007). 'Evolution in the hypervariable environment of Madagascar.' *PNAS* 104: 13723–7.

Dirac, C., L. Andriambelo et al. (2006). 'Scientific bases for a participatory forest landscape management in Central Menabé.' *Madagascar Cons. and Devel.* 1: 31–3.

Dolins, F., A. Jolly et al. (2010). 'Conservation education in Madagascar: three case studies of an island-continent.' *Amer. J. Primatol* 72: 391–406.

Eisenberg, J.F. (1981). *The Mammalian Radiations.* Chicago: University of Chicago Press.

Evers, S.J.T.M., C. Campbell et al. (2013). 'Land competition and human–environment relations in Madagascar.' In S.J.T.M. Evers, C. Campbell and M. Lambek (eds), *Contest for Land in Madagascar.* Leiden: Brill, pp. 3–20.

Falloux, F., and L.M. Talbot. (1993). *Crisis and Opportunity: Environment and Development in Africa*. London: Earthscan.
Fedigan, L.M. (1994). 'Science and the successful female: why there are so many women primatologists.' *American Anthropologist* 96: 529–40.
Fedigan, L.M. (1997). 'Is primatology a feminist science?' In L. Hager (ed.), *Women in Human Evolution*. New York: Routledge, pp. 55–75.
Fellman, D. (2014). *Island of Lemurs: Madagascar.* Montreal: IMAX.
Flacourt, É. de. (1661). *Histoire de la Grande Ile de Madagascar.* Troyes: Nicolas Oudot.
Ford, T. (2013). 'Madagascar's trees vanish to feed taste for rosewood in west and China.' *Guardian*, December 23: 1.
Frère, S. (1958). *Panorama de l'Androy.* Paris: Éditions Aframpe.
Freudenberger, K. (2010). 'Paradise lost? Lessons from 25 years of USAID environment programs in Madagascar.' Washington DC: International Resources Group, USAID.
Freudenberger, M.S., and K. Freudenberger. (2002). 'Contradictions in agricultural intensification and improved natural resource management: issues in the Fianarantsoa forest corridor of Madagascar.' In C.B. Barret, F. Place and A.A. Aboud (eds), *Natural Resources Management in Tropical Agriculture: Understanding and Improving Current Practices*. Oxford, CAB Publishing.
Ganzhorn, J.U. (1995). 'Cyclones over Madagascar: fate or fortune?' *Ambio* 24: 124–5.
Ganzhorn, J.U., S.M. Goodman et al. (2007). *Biodiversity, Ecology and Conservation of Littoral Ecosystems in Southeastern Madagascar, Tolagnaro (Fort Dauphin)*. Washington DC: Smithsonian Institution.
Gautier, L., Y. Naciri et al. (2013). 'A new species, genus and tribe of Saptaceae, endemic to Madagascar.' *Taxon* 62: 972–83.
Geertz, C. (1963). *Agricultural Involution: The Process of Ecological Change in Indonesia*. Berkeley: University of California Press.
Golden, C.D., M.J. Bonds et al. (2013). 'Economic valuation of subsistence harvest of wildlife in Madagascar.' *Conservation Biology* 28(1).
Goodman, S.M., and J.P. Benstead (eds). (2003). *The Natural History of Madagascar.* Chicago: Chicago University Press.
Gordon, A.D., S.E. Johnson et al. (2013). 'Females are the ecological sex: sex-specific body mass ecogeography in wild sifaka populations (*Propithecus* spp.).' *Amer. J. Phys Anthrop.* 151: 77–87.
Gould, L., R.W. Sussman et al. (1999). 'Natural disasters and primate populations: the effects of a two-year-drought on a naturally occurring population of ringtailed lemurs (*Lemur catta*) in southwestern Madagascar.' *Int. J. Primatol.* 20: 69–85.
Gould, L., R.W. Sussman et al. (2003). 'Demographic and life-history patterns in a population of ring-tailed lemurs (*Lemur catta*) at Bezà

Mahafaly Reserve, Madagascar: a 15-year perspective.' *Amer. J. Phys Anthrop.* 120: 182–94.

Grubb, P.J. (2003). 'Interpreting some outstanding features of the flora and vegetation of Madagascar.' *Perspect. Plant Ecol. Evol. Syst.* 6: 125–46.

Gwinnett, G. (2013). 'Madagascar Oil gears up for an exciting six months.' September 4. www.proactiveinvestors.com/companies/news/47694/madagascar-oil-gears-up-for-exciting-six-months-47694.html.

Hannah, L., R. Dave et al. (2008). 'Climate change adaptation for conservation in Madagascar.' *Biology letters* 4: 590–94.

Hanson, P.W. (2009). 'Engaging green governmentality through ritual.' *Études Océan Indien* 42–43: 1–26.

Hanson, P.W. (2012). 'Toward a more transformative participation in the conservation of Madagascar's natural resources.' *Geoforum* 43(6).

Haraway, D.J. (1989). *Primate Visions: Gender, Race, and Nature in the World of Modern Science.* New York: Routledge.

Harbinson, R. (2007) Development recast: a review of the impact of Rio Tinto ilmenite mine in Madagascar: The impact of the Rio Tinto ilmenite mine in Madagascar.' Friends of the Earth.

Harper, J. (2002). *Endangered Species: Health, Illness and Death among Madagascar's People of the Forest.* Durham NC: Carolina Academic Press.

Harper, J. (2008). 'The environment of environmentalism: turning the ethnographic lens on a conservation project.' In J.C. Kaufmann (ed.), *Greening the Great Red Island: Madagascar in Nature and Culture.* Pretoria: Africa institute of South Africa, pp. 241–74.

Hrdy, S., and P. Wright (2014). 'Alison Jolly: A supremely social intelligence (1937–2014).' *Evolutionary Anthropology* 23: 121–5.

Hutchinson, G.E. (1965). *The Ecological Theater and the Evolutionary Play.* New Haven CT: Yale University Press.

ICMM. (2013). 'ICMM: International Council on Mining and Metals.' www.icmm.com.

Jolly, A. (1966). *Lemur Behavior.* Chicago: University of Chicago Press.

Jolly, A. (1966). 'Lemur social behavior and primate intelligence.' *Science* 153: 501–6.

Jolly, A. (1972). *The Evolution of Primate Behavior.* New York: Macmillan.

Jolly, A. (1980). *A World Like Our Own: Man and Nature in Madagascar.* New Haven CT: Yale University Press.

Jolly, A. (1984). 'The puzzle of female feeding priority.' In M.F. Small (ed.), *Female Primates: Studies by Women Primatologists.* New York: Liss, pp. 197–205.

Jolly, A. (1987). 'Madagascar, a world apart.' *National Geographic* 171: 148–83.
Jolly, A. (1988). 'Madagascar lemurs: on the edge of survival.' *National Geographic* 174: 132–61.
Jolly, A. (1991). 'Female biology and women biologists.' *Trends in Ecology and Evolution* 6: 39–40.
Jolly, A. (1999). *Lucy's Legacy: Sex and Intelligence in Human Evolution*. Cambridge MA: Harvard University Press.
Jolly, A. (2004). *Lords and Lemurs: Mad Scientists, Kings with Spears, and the Survival of Diversity in Madagascar*. Boston MA: Houghton Mifflin.
Jolly, A. (2011). 'The narrator's stance: story-telling and science at Berenty Reserve.' In J. MacClancy and A. Fuentes (eds), *Centralizing Fieldwork: Critical Perspectives from Primatology, Biological and Social Anthropology*. New York: Berghan Books, pp. 223–42.
Jolly, A. (2012). 'Berenty Reserve, Madagascar: A long time in a small space.' In P. Kappeler and D. P. Watts (eds), *Long-Term Field Studies of Primates*. Heidelberg: Springer: 29–44.
Jolly, A., S. Caless et al. (2000). 'Infant killing, wounding, and predation in *Eulemur* and *Lemur*.' *Int. J. Primatol.* 21: 21–40.
Jolly, A., and M. Jolly (1990). 'A view from the other end of the telescope.' *New Scientist* 58.
Jolly, A., P. Oberlé et al. (eds). (1984). *Madagascar*. Key Environments. Oxford: Pergamon.
Jolly, A., and R.E. Pride. (1999). 'Troop histories and range inertia of *Lemur catta* at Berenty: a 33 year perspective.' *Int. J. Primatol.* 20: 359–73.
Jolly, A., H. Rasamimanana et al. (2006). 'Territory as bet-hedging: *Lemur catta* in a rich forest and an erratic climate.' In A. Jolly, R.W. Sussman, N. Koyama and H. Rasamimanana (eds), *Ringtailed Lemur Biology: Lemur catta in Madagascar*. New York, Springer, pp. 187–207.
Jolly, A., H. Rasamimanana and D. Ross. (2005–2012). *Ako the Aye-Aye, Bitika the Mouselemur, Tik-Tik the Ringtailed Lemur, Bounce the White Sifaka, Furry and Fuzzy the Red Ruffed Lemur Twins, No-Song the Indri*. The Ako Series. Myakka City FL: Lemur Conservation Foundation.
Jolly, M., A. Jolly and H. Rasamimanana (2010). 'The story of a friendship.' *Madagascar Conservation & Development* 5: 125–6.
Jolly, R. (2014). *UNICEF: Global Governance that Works*. Abingdon: Routledge.
Jolly, R. (ed.). (2001). *Jim Grant, UNICEF Visionary*. Florence: Unicef.
Keller, E. (2008). 'The banana plant and the moon: conservation and

the Malagasy ethos of life in Masoala, Madagascar.' *American Ethnologist* 35: 650–54.

Kull, C. (1996). 'The evolution of conservation efforts in Madagascar.' *International Environmental Affairs* 3(1) (Winter): 50–86.

Kull, C. (2004). *Isle of Fire: The Political Ecology of Landscape Burning in Madagascar.* Chicago: University of Chicago Press.

Labastille, A. (1983). 'Eight women in the wild.' *International Wildlife* 13: 36–43.

Lanting, F., A. Jolly and J. Mack (1990). *Madagascar: A World Out of Time.* New York: Aperture.

Lührs, M.-L. (2013). 'Simultaneous GPS tracking reveals male associations in a solitary carnivore.' *Behav. Ecol. and Sociobiol.* 67: 1731–43.

Lührs, M.-L., M. Dammhahn et al. (2013). 'Strength in numbers: males in a carnivore grow bigger when they associate and hunt cooperatively.' *Behav. Ecol.* 24.

Lührs, M.-L., and Peter M. Kappeler (2014). 'Polyandrous mating in treetops: how male competition and female choice interact to determine an unusual carnivore mating system.' *Behav. Ecol. and Sociobiol.* 68(6): 879–89.

Madagascar Oil. (2013). 'Steaming towards first oil.' November 26. www.madagascaroil.com/documents/MadagascarOilAfrica Oil-Week STFO v2.pdf.

Martin, R.D. (2013). *How We Do It: The Evolution and Future of Human Reproduction.* New York: Basic Books.

MEEF (1984). *Stratégie malgache pour la conservation et le développement durable.* Elaboré par la Commission National de la Conservation des ressources vivantes au service du développement national insitué par le décret présidentiel No. 84–116 du 4 avril 1994 et avec l'assistance de UICN/WWF. Antananarivo.

Meier, B., R. Albignac et al. (1987). 'A new species of *Hapalemur* (Primates) from South East Madagascar.' *Folia primatol.* 48: 211–15.

Mercier, J.-R. (2006). 'The preparation of the National Environmental Action Plan (NEAP): was it a false start?' *Madagascar Conservation and Development* 1: 50–54.

Middleton, K. (2000). 'The rights and wrongs of loin-washing.' *Taloha* 13: Repenser la femme malgache: de nouvelles perspectives sur le genre à Madagascar: 63–99.

Missouri Botanical Garden. (1995–2013). 'Using aerial photography to measure Madagascar's littoral forests.' www.mobot.org/MOBOT/research/littoral.

Myers, N., R.A. Mittermeier et al. (2000). 'Biodiversity hotspots for conservation priorities.' *Nature* 403: 853–8.

Perrier de la Bathie, H. (1921). *La Végétation Malgache*. Marseilles: Annales du Musée Colonial de Marseille.
Peters, J. (1997). 'Local participation in the conservation of the Ranomafana National Park, Madagascar.' *J. World Forest Resources Management* 8: 109–35.
Peters, J. (1998). 'Sharing national park entrance fees: forging new partnerships in Madagascar.' *Society and Natural Resources* 11: 517–30.
Peters, J. (1998). 'Transforming the Integrated Conservation and Development Project approach: observations from the Ranomafana National Park Project, Madagascar.' *J. Agricult. Envir. Ethics* 11: 17–47.
Peters, J. (1999). 'Understanding conflicts between people and parks at Ranomafana, Madagascar.' *Agriculture and Human Values* 16: 65–74.
Petter, J.-J. (1962). 'Recherches sur l'écologie et l'éthologie des lémuriens malgaches.' *Mém. Mus. Natl. Hist. Nat.* (Paris) 27: 1–146.
Petter-Rousseaux, A. (1962). 'Recherches sur la Biologie de la Réproduction des Primates Inférieurs.' *Mammalia* 3794: 1–87.
Piccirilli, J. (2010). *The Art and Life of Alison Mason Kingsbury*. Ithaca NY: Cornell University Library.
Pochron, S.T., J. Fitzgerald et al. (2003). 'Patterns of female dominance in *Propithecus diadema edwardsi* of Ranomafana National Park, Madagascar.' *Amer. J. Primatol.* 61: 173–85.
Pochron, S.T., W.T. Tucker et al. (2004). 'Demography, life history and social structure of *Propithecus diadema edwardsi* from 1986–2000 in Ranomafana National Park, Madagascar.' *Amer. J. Phys. Anthropol.* 125: 61–72.
Pollini, J. (2007). 'Slash-and-burn cultivation and deforestation in the Malagasy rainforests: representations and realities.' Ph.D. thesis, Department of Natural Resources, Cornell University, p. 776.
Pollini, J. (2011). 'The difficult reconciliation of conservation and development objectives: the case of the Malagasy Environmental Action Plan.' *Human Organization* 70: 74–87.
Pride, R.E., D. Felantsoa et al. (2006). 'Resource defence in *Lemur catta:* the importance of group size.' In A. Jolly, N. Koyama, H.R. Rasamimanana and R.W. Sussman (eds), *Ringtailed Lemur Biology: Lemur catta in Madagascar.* New York: Springer, pp. 208–32.
Raharimanana, J.-L.V., and C. Ravoajanahary (2007). *Madagascar, 1947.* La Roque d'Antheron: Vents d'ailleurs.
Ralaimihoatra, E.D. (1966). *Histoire de Madagascar.* Tananarive: Société malgache d'édition.
Ramahatra, R., and H. Patterson. (1993). *Poverty in Madagascar.* Antananarivo: UNDP.

Randriamahazo, H. (2011). 'Radiated tortoises and the fading taboo.' *Turtle Survival*, August: 63–8.

Raonimanalina, N., and W. Fitzgibbon. (2014). 'Chinese oligarch could face scrutiny in Madagascar oil land grab.' *Africa Report*, January 8.

Rasamimanana, H.R., V.N. Andrianome et al. (2006). 'Male and female ringtailed lemurs' energetic strategy does not explain female dominance.' In A. Jolly, N. Koyama, H.R. Rasamimanana and R.W. Sussman (eds), *Ringtailed Lemur Biology: Lemur catta in Madagascar*. New York: Springer.

Rasoloarison, R.M., S.M. Goodman et al. (2000). 'Taxonomic revision of mouse lemurs *(Microcebus)* in the western portions of Madagascar.' *Int. J. Primatol.* 21: 963–1020.

Ratovomamana, R.Y., C. Rajeriarison et al. (2011). 'Phenology of different vegetation types in Tsimanampetsotsa National Park, southwestern Madagascar.' *Malagasy Nature* 5: 14–38.

Ratsimbazafy, J., L.J. Rakotoniaina et al. (2008). 'Cultural anthropologists and conservationists: can we learn from each other to conserve the diversity of Malagasy species and culture? In J.C. Kaufmann (ed.), *Greening the Great Red Island: Madagascar in Nature and Culture*. Pretoria: Africa Institute of South Africa, pp. 301–15.

Ratsirarson, J., and A.F. Richard. (2013). *Interim Report on Bezà Mahafaly to the Liz Claiborne Art Ortenberg Foundation*. Antananarivo: Liz Claiborne Art Ortenberg Foundation.

Raxworthy, C.J., R.G. Pearson et al. (2008). 'Extinction vulnerability of tropical montane endemism from warming and upslope displacement: a preliminary appraisal from the highest massif in Madagascar.' *Global Change Biology* 14: 1703–20.

Razafindrakoto, M., F. Roubaud et al. (2013). *Institution, gouvernance et croissance de long terme à Madagascar: l'enigme et le paradox*. Marseilles: Institut de Recherche pour le Développement, pp. 1–35.

Richard, A.F., R.E. Dewar et al. (2002). 'Life in the slow lane? Demography and life histories of male and female sifaka (*Propithecus verreauxi verreauxi*).' *J. Zool., Lond.* 256: 421–36.

Richard, A.F., and J. Ratsirarson. (2013). 'Partnership in practice: making conservation work at Bezà Mahafaly, southwest Madagascar.' *Madagascar Cons. and Devel.* 8: 12–20 (www.bezamahafaly.commons.yale.edu).

Rio Tinto. (2013). 'Rio Tinto QIT Madagascar Minerals.' www.riotinto-madagascar.com/english/index.asp.

Sauther, M.L. (1993). 'Resource competition in wild populations of ringtailed lemurs (*Lemur catta*): implications for female dominance.' In P.M. Kappeler and J.U. Ganzhorn (eds), *Lemur Social Systems and their Ecological Basis*. New York: Plenum, pp. 135–52.

Sauther, M.L., F. Cuozzo et al. (2013). 'Limestone cliff-face and cave use by wild ring-tailed lemurs (*Lemur catta*) in southwestern Madagascar.' *Madagascar Cons. and Devel.* 8: 75–80.

Sauther, M.L., R.W. Sussman et al. (1999). 'The socioecology of the ringtailed lemur: thirty-five years of research.' *Evol. Anthropol.* 8: 120–32.

Scardina, J. (2012). *Wildlife Heroes: 40 Leading Conservationists and the Animals They are Committed to Saving.* Philadelphia PA: Running Press.

Schwitzer, C., C.G. Mittermeier et al. (2013). *Lemurs of Madagascar: A Strategy for Their Conservation 2013–2016.* Bristol: IUCN SSC Primate Specialist Group, Bristol Conservation and Science Foundation and Conservation International.

Seagle, C. (2012). 'Inverting the impacts: mining, conservation and sustainability claims near the Rio Tinto/QMM ilmenite mine in Southeast Madagascar.' *J. Peasant Studies* 39(2).

Sen, A. (1981). *Poverty and Famines: An Essay on Entitlement and Deprivation.* Oxford: Clarendon Press.

Shapiro, B., A. Woldeyes et al. (2010). *Nourishing the Land, Nourishing the People: The Story of One Rural Development Project in the Deep South of Madagascar that Made a Difference.* Rome and Cambridge MA: IFAD and COBI.

Shyamsundar, P., and R.A. Kramer. (1996). 'Tropical forest protection: an empirical analysis of the costs born by local people.' *J. Envir. Econ. and Management* 32: 129–44.

Shyamsundar, P., and R.A. Kramer. (1997). 'Biodiversity conservation—at what cost? A study of households in the vicinity of Mantadia National Park.' *Ambio* 26: 180–84.

Simmen, B., F. Bayart et al. (2010). 'Total energy expenditure and body composition in two free-living sympatric lemurs.' *PLosONE.* 55(3): 1–10.

Sodikoff, G.M. (2012). *Forest and Labor in Madagascar: From Colonial Concession to Global Biosphere.* Bloomington: Indiana University Press.

Sodikoff, G.M. (2012). 'Totem and Taboo reconsidered: Endangered species and moral practice in Madagascar (Mananara).' In G.M. Sodicoff (ed.), *Anthropology of Extinction: Essays on Culture and Species Death.* Bloomington IN, Indiana University Press.

Sorg, J.-P., J.U. Ganzhorn et al. (2003). 'Forestry and research in the Kirindy Forest Center/Centre de Formation Professionnelle Forestière.' In S.M. Goodman and J.M. Benstead (eds), *The Natural History of Madagascar.* Chicago: University of Chicago Press, pp. 1512–19.

Sterling, E.J., and E.E. McCreless. (2006). 'Adaptations in the aye-aye: a review.' In L. Gould and M.L. Sauther (eds), *Lemurs: Ecology and Adaptation.* New York: Springer, pp. 159–84.

SuLaMa (2011). *Diagnostic participatif de la gestion des ressources naturelles sur le plateau Mahafaly.* University of Antananarivo and University of Hamburg.

Summerhill, M. (2011). *Madagascar: Land of Heat and Dust*, BBC Earth. Broadcast February 23.

Sussman, R.W., A.F. Richard et al. (2012). 'Bezà Mahfaly Special Reserve: long-term research on lemurs in southwestern Madagascar.' In P. Kappeler and D.P. Watts (eds), *Long-Term Field Studies of Primates*. Heidelberg: Springer Verlag, pp. 45–66.

Tadross, M., L. Randriamarolaza et al. (2008). *Climate Change in Madagascar: Recent Past and Future.* Washington DC and Cape Town: World Bank and Climate Systems Analysis Group, University of Cape Town.

Thomson, S. (2011). *David Attenborough and the Giant Egg.* BBC. Broadcast March 2.

Tollefson, J. (2013). 'Splinters of the Amazon.' *Nature* 496: 286–9.

UNESCO Biosphere Reserve Information: Madagascar, Mananara-Nord, UNESCO.

USAID. (2013). *Nature, Wealth and Power: Leveraging Natural and Social Capital for Resilient Development.* Washington DC: USAID.

ValBio (2013). 'ICTE–Centre ValBio publications.' www.stonybrook.edu/commcms/centre-valbio/research/publications.html.

Vick, L.G., and M.E. Pereira. (1989). 'Episodic targeted aggression and the histories of lemur social groups.' *Behav. Ecol. Sociobiol.* 25: 3–12.

Virah-Swamy, M. (2009). 'Threshold response of Madagascar's littoral forest to sea-level rise.' *Global Ecol. and Biogeography* 18: 96–100.

Walker, R.C., and T.H. Rafeliarisoa. (2011). 'The precarious conservation status of the critically endangered Madagascar spider tortoise (*Pyxis arachnoides*) and radiated tortoise (*Astrochelys radiata*): what we now know through three years of field operations.' *Turtle Survival*, August: 69–74.

Wells, N.A. (2003). 'Some hypotheses on the Mesozoic and Cenozoic paleoenvironmental history of Madagascar.' In S.M. Goodman and J.P. Benstead (eds), *The Natural History of Madagascar.* Chicago: University of Chicago Press, pp. 16–40.

Wildife Conservation Society. (2013). 'Madagascar puts first-ever government-backed carbon credits on open market.' www.wcs.org/press/press-releases/makira-carbon-sale.aspx.

World Bank. (1988). *Madagascar: Plan d'Action Environnemental*, Volume 1: *Document de synthèse générale et propositions d'orientations.* Washington DC and Antananarivo: World Bank.

World Bank. (2013). *Madagascar Country Environmental Analysis.* Washington DC: World Bank.

Wright, P.C. (1999). 'Lemur traits and Madagascar ecology: coping with an island environment.' *Yrb. Phys. Anthropol.* 42: 31–72.

Wright, P.C., and B.A. Andriamihaja (2002). 'Making a rain forest national park work in Madagascar: Ranomafana National Park and its long-term research commitment.' In J. Terborgh, C. vanSchaik, M. Rao and L. Davenport (eds), *Making Parks Work: Strategies for Preserving Tropical Nature*. Covelo CA: Island Press, pp. 112–36.

Wright, P.C., and B.A. Andriamihaja. (2003). 'The conservation value of long-term research: a case study of the Parc National de Ranomafana.' In S. Goodman and J. M. Benstead (eds), *The Natural History of Madagascar*. Chicago: Chicago University Press.

Wright, P.M. (1995). 'Demography and life history of free-ranging *Propithecus diadema edwardsi* at Ranomafana National Park, Madagascar.' *Int. J. Primatol.* 16: 835–53.

Photographic credits

© Frans Lanting: Alison Jolly climbing into a boat; Rainforest, Ranomafana National Park; Deforested hills; Antananarivo; Sportive lemur in tree hole; Logger felling tree in rainforest; Man cutting baobab bark; Antandroy leading cattle to market; Rekanoky, chief reserve guardian at Berenty; Parson's chameleon male.

© Cyril Ruoso: Cover photo; Berenty Reserve at night; Zebu cattle drove; Baobab tree; Sportive lemur; Verreaux's sifakas in trees; Verreaux's sifaka leaping; Mother and baby ring-tailed lemur; Mother and baby ring-tailed lemurs playing in trees; Malagasy woman holding a baby ring-tail; Advance of a troop of ring-tails; Gobe mouche; Radiated tortoise.

© Noel Rowe: Looking out from Namanabe Hall; Waterfall in Ranomafana National Park; *Daubentonia madagascariensis*; *Hapalemur aureus*; Indri; Fosa eating a pigeon.

© Russell Mittermeier: Madame Berthe's mouse lemur.

© Sheila M. O'Connor: Duke of Edinburgh.

© Ando Ratovonirina: Hanta and Alison at Lake Alaotra.

© Derek Oayes: Alison on the cover of *Primate Behavior.*

David Attenborough and Alison Jolly: photo taken during the BBC filming of their natural history series on Madagascar, 2010.

All other photos are © Jolly family archive.

Index